Student Study Guide and Solutions Manual

for

Wink, Fetzer-Gislason, and McNicholas's

THE PRACTICE OF CHEMISTRY

Pamela Mills
Hunter College
Amina K. El-Ashmawy
Collin County Community College

W. H. Freeman and Company
New York

ISBN-13: 978-0-7167-4875-5
ISBN-10: 0-7167-4875-4

Printed in the United States of America

Second Printing

W. H. Freeman and Company
41 Madison Avenue
New York, NY 10010
Houndmills, Basingstoke
RG21 6XS, England
www.whfreeman.com

Table of Contents

It is with great love, admiration, respect, and gratitude that I dedicate this work to my role model, my mentor, my inspiration—Dr. Aida M. Geumei, my mother.

—Amina K. El-Ashmawy

CHAPTER 1

Elements and Compounds: The Chemist's View of Matter

Checklist

What you need to be able to do when you finish Chapter 1

- Be able to classify matter into elements and compounds and into pure substances and mixtures.

  Example: Keep in mind that some substances may fall into two categories. For example, elements are also pure substances. Identify the element, compound, pure substance, and mixture from the following list: Water, air, nitrogen.

- Know the difference between an atom and a molecule.
- Become familiar with the periodic table. Know the names and symbols of the first 20 elements.
- Look at a chemical formula, identify the atoms in the formula, and find each atom in the periodic table. Using the table, give the group number and period number for the atom.
- Know all about the atom and its symbol. Given an atom, how many protons, neutrons, and electrons does the atom have? What is its atomic number? What is its mass number?
- Be able to write atomic symbols for an atom using the periodic table given the number of neutrons in the atom.
- Distinguish between physical and chemical properties of matter.
- Identify the three states of matter and the names of the processes that cause matter to change state.
- Look at the periodic table again and identify the categories that chemists use and the elements that belong in the categories. Identify the elements that are metals, semimetals, and nonmetals. Identify the halogens, the chalcogens, the alkali metals, and the alkaline earth metals.

Practice Problems and Study Hints

Practice 1 Categorizing and More Categorizing

Scientists love to categorize things. By organizing objects, processes, and ideas into categories, it is easier to understand and retain the information. Chemists are masters at categorizing. The periodic table contains a wealth of information all organized into categories. Let's first organize matter according to the following diagram:

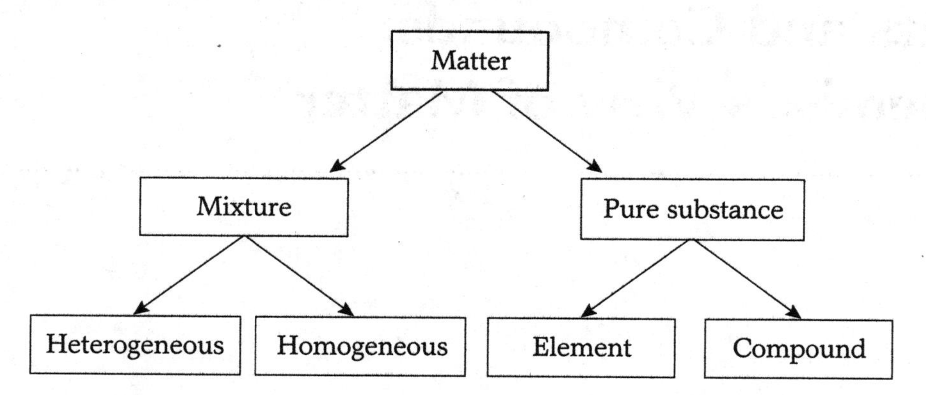

Look at the diagram above. Notice how we start with something completely uncharacterized (matter) and then categorize it into a specific description.

Using the chart, write a one-sentence description of each category in the chart. Give an example, drawn from your common experience, of each of the final four categories (heterogeneous and homogeneous mixture, and element and compound).

Practice 2 Categorizing and Simplifying

Another way chemists categorize is by thinking about matter on a macroscopic scale ("This water is clear, wet, and heavy.") and on a molecular—or atomic—scale ("Is water a molecule? What are its constituent atoms?"). We need to think both macroscopically (as we did in Practice 1) and microscopically. To think on a microscopic scale, we need to distinguish between atoms and molecules.

There is a connection between the micro and macroscopic scales. For example, an atom is to an element what a molecule is to a compound. Describe this relationship. How are atoms and elements related? How are they different? How are molecules and compounds related? How are they different?

Practice 3 The Periodic Table: The Chemists Grand Categorizing Scheme

The periodic table categorizes elements in a variety of ways. Look at a periodic table and test your knowledge of it.

1. First, write down the names of the first 20 elements. How many do you know? How many did you get wrong? Make a list of the elements you do not know and keep memorizing them until you know them all. Check the spelling to be sure each element is spelled correctly.

2. Identify the metals, nonmetals, and semimetals on the table. Categorize the following as metals, semimetals, or nonmetals: O, Hg, U, Cr, Kr, Rb, Na, Ga, Al, B.
3. Close your eyes and pick an atom from the periodic table. Identify the atom's group number and its period number. Do this until you are sure you will always get it right. Remember, you can always refer to the periodic table.
4. Identify the group of alkali metals on the periodic table. Identify the alkaline earth metals. Identify the halogens, the inert gases, and the chalcogens. Using the table, give the symbols for the inert gas, alkali metal, alkaline earth metal, halogen, and chalcogen that correspond to period four.
5. Identify the transition metals. Identify the inner transition elements. Pick one of the transition metals, and identify its symbol, group number, and period number. Pick a representative inner transition element and identify its symbol, group number, and period number.

Advice: You should master the above exercises. You can do the exercises with the periodic table in front of you, but you should understand its categories and make no mistakes in categorizing the atoms. (OK, you can make one mistake—but no more!)

Practice 4 The Quantitative Aspects of the Periodic Table

Look at the periodic table and be sure you know, in general, where the atomic number of each element appears in the table. What does the atomic number mean?

Pick any atom from the table. For example, pick Ag. What is its atomic number? How many protons does the Ag atom have? (Remember that the atomic number always tells you how many protons the atom has. In fact, the number of protons *defines* the atom type.) How many electrons does the neutral Ag atom have?

Pick another atom, and identify the number of protons and electrons the neutral atom has. Do this until you feel that you are certain you can find the number of protons for a given atom from the periodic table.

Practice 5 The Particles of the Atom

We will leave the periodic table now and focus on the atom. What are the constituent particles of the atom? What are the relative charges of the particles? Which particles have the greater mass?

Write a one-paragraph description of the atom. Imagine that you are describing the atom to your friend who knows absolutely nothing about chemistry. Include in your description the three types of particles of the atom, where they are located (nucleus, core, valence shell), their relative masses and charges, and which particle determines the type of the atom. Also include a sketch of an atom to help further illustrate the concept of an atom for your friend.

Practice 6 The Mass Number and Atomic Symbols

The mass number of an atom is simply the sum of the protons and neutrons of the atom. Unfortunately, we cannot read this number from the periodic table. We can only determine the number of neutrons one of two ways: by the number or from the atomic symbol. If we are given the number of neutrons, we should be able to write the atomic symbol.

Complete the following table (which is similar to Example 1.9) using everything you know about the periodic table, mass number, and/or atomic symbols. Assume all the atoms in the table are neutral.

Element	Atomic Number	Number of Protons	Number of Neutrons	Number of Electrons	Atomic Number	Mass Number
Phosphorous	$^{31}_{15}P$	15	16	15	15	31
Cadmium	$^{112}_{48}Cd$	48	64	48	48	112
Iron	$^{56}_{26}Fe$	26	30	26	26	56
Yttrium	$^{89}_{39}Y$	39	50	39	39	89
Lithium	$^{7}_{3}Li$	3	4	3	3	7
	$^{19}_{9}F$	9	10	9	9	19

Practice 7 Physical Properties of the Elements

1. Pick a metal with which you are familiar (such as copper or iron). List three physical properties of the metal. At room temperature, is it a solid, liquid, or gas? Suppose you heat the metal to really high temperatures. How will it change its state? What is the name of this process?
2. Pick an element with which you are familiar that is a gas at room temperature. List three physical properties of this gas. Imagine that the gas is cooled until it liquefies. What is the name of this process?
3. We are most familiar with substances that change their physical state in a particular order. Ice, when heated, becomes water and then steam. If we start with a solid we imagine that we can heat it until it becomes a liquid and then a gas. Changing a solid to a liquid is the process of melt ; changing a liquid to a gas is the process of evaporating . Sometimes, we can change a solid to a gas. Carbon dioxide is the classic example. This is the process of sublimation .
4. Examine Figure 1.17, and write an example for all the processes listed there if you can. You may not know of an example of depositing a gas. See if anyone in your study group has an example.

CHAPTER 2

Molecular Substances and Lewis Structures

Checklist

What you need to be able to do when you finish Chapter 2

- From the chemical formula, identify whether or not the compound is molecular.

  Example: Which of the following compounds are molecular: NO_3, $RbNO_3$, MgO?

- Know the seven diatomic elements and know their names (page 49).
- Name binary molecular compounds from the chemical formula and be able to construct the chemical formula from the name of the compound.

  Example: What is the name of the compound SF_6? What is the chemical formula of dinitrogen tetroxide?

- Identify possible gases at room temperature from the chemical formula.

  Example: Which of the following compounds are likely to be gases at room temperature: NO, CS_2, P_2O_5, MgO?

- Draw Lewis structures from the chemical formula for molecular compounds and ions.

  Example: Draw the Lewis structure for BrO_2^-.

- In cases where more than one valid Lewis structure is possible, draw the possible resonance structures.

  Example: What are all the resonance structures for NO_2^-?

Organizing Our Knowledge

Molecular Substances, Elements, and Compounds

Let's construct a flow diagram of our knowledge from this chapter beginning with the basic information learned in Section 2.1 about molecules and their names.

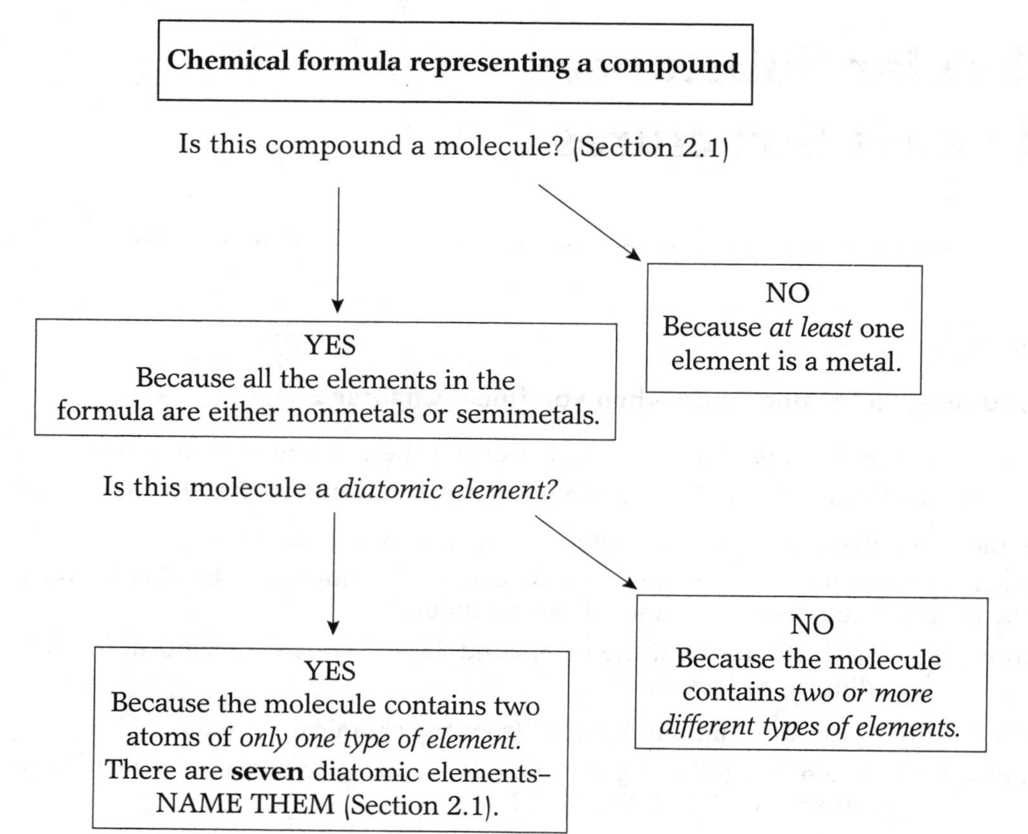
Chemical formula representing a compound
Is this compound a molecule? (Section 2.1)
YES
Because all the elements in the formula are either nonmetals or semimetals.
NO
Because *at least* one element is a metal.
Is this molecule a *diatomic element?*
YES
Because the molecule contains two atoms of *only one type of element.* There are **seven** diatomic elements– NAME THEM (Section 2.1).
NO
Because the molecule contains *two or more different types of elements.*

Practice Problems and Study Hints

Practice 1 Recognizing Elements and Compounds

It is essential that you master the above diagram and can answer questions such as the following.

- Indicate whether or not the following compounds are likely to be molecular:
  a. PCl_5
  b. CsCl
  c. NOCl
- Name the seven diatomic elements. (See page 49.)
- When one refers to oxygen, what *chemical formula* is meant, O or O_2? (See page 49.)
- Give the *names* of the seven diatomic elements.

H_2, O_2, N_2, Cl_2, I_2, F_2, Br_2

Naming Molecules

It is important to realize that we are going to name one type of chemical compound—molecules. So if, in the previous diagram, you answered NO to the question "Is the compound a molecule," you cannot yet give the systematic name of the compound. Only if you answer YES to that question can you name the molecule. There are even more limitations. Let's pick up the diagram and add more to it:

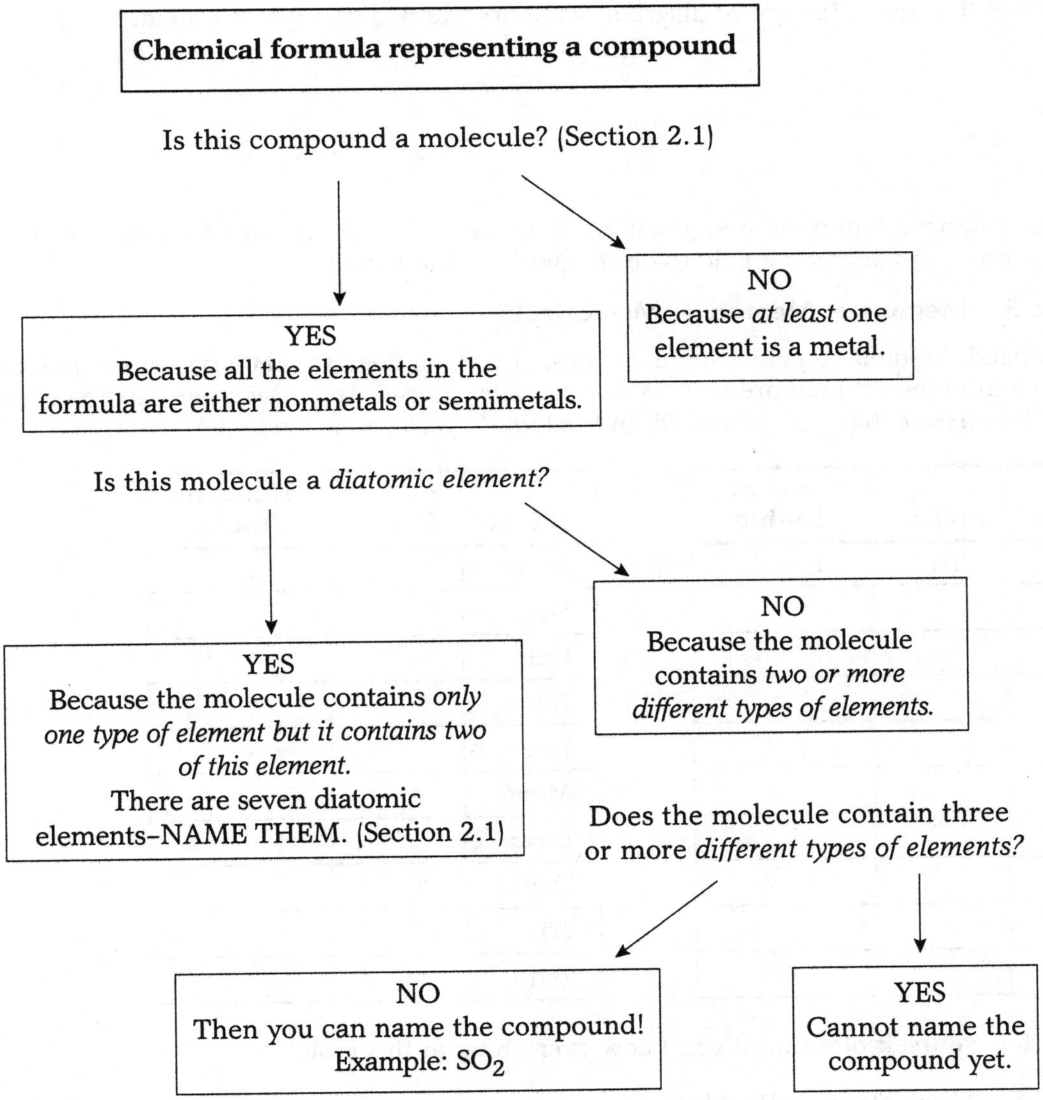

Chemical formula representing a compound

Is this compound a molecule? (Section 2.1)

YES
Because all the elements in the formula are either nonmetals or semimetals.

NO
Because *at least* one element is a metal.

Is this molecule a *diatomic element?*

YES
Because the molecule contains *only one type of element but it contains two of this element.*
There are seven diatomic elements–NAME THEM. (Section 2.1)

NO
Because the molecule contains *two or more different types of elements.*

Does the molecule contain three or more *different types of elements?*

NO
Then you can name the compound!
Example: SO_2

YES
Cannot name the compound yet.

Naming Practice

Before doing any naming, look at the above diagram and decide which compounds you should be able to name.

Practice 2 Naming Compounds

Of the following four compounds, you should be able to name two based on the diagram we have just used. You might be able to name more than that because you have additional information. But using the above diagram, what are the two that you can name?

a. CsF

b. N_2O_5

c. CF_4

d. NH_4NO_3

Now that you have found the two, what are their names? (Refer to page 51 if you need help. But remember, you should be able to do this without the table.)

Practice 3 Memorize, Memorize, Memorize!

Did you need the table to name the molecules? If so, you need to memorize those prefixes. First, do a mini-test. Which prefixes do you *absolutely know*? And which ones give you trouble? To answer these questions, fill in the table. I will do the first one.

Count	Prefix	Know or Trouble
3	Tri	Know
7	hepta	
1	mono	
6	hexa	
10	deca	
2	di	
5	penta	
8	octa	
4	tetra	
9	nona	

Prefix	Count	Know or Trouble
Penta	5	
Di	2	
Octa	8	
Nona	9	
Deca	10	
Mono	1	
Hexa	6	
Hepta	7	
Tri	3	
Tetra	4	

Advice: Test yourself often until you know everything in this table!

Practice 4 More Naming Problems

1. Name the following molecules:

   a. P_4O_6 —tetra phosphorus hexoxyde

   b. NBr_3 – Nitrogen tribromide

   c. S_2F_2 disulfur difluoride

2. Give the formula for the following molecules from the name:
   a. Dinitrogen trioxide N_2O_3
   b. Iodine heptafluoride IF_5
   c. Chlorine trifluoride ClF_3

Practice 5 Make Up Your Own Molecules

If you want more practice, then make up your own molecules. For right now, don't worry about whether or not the molecule really exists—just make it up and name it. Or name it and then write the formula. Remember the rules: 1) it must be a molecule, and 2) it must be *binary.*

For example, here is a nonsense molecule: P_5S_2. It would have the name pentaphosphorous disulfide. As a chemistry professor, I should not be naming phony molecules. But that shouldn't stop you. Play this game until you are *completely proficient at naming molecules.*

Practice 6 What Are Gases?

Reread Practical "A." You should now realize that you can use the molecular formula to tell which compounds are gases. Write the chemical formula for the following compounds and indicate whether or not the compound is likely to be a gas at room temperature. Explain *why* you selected those as gases and why you rejected the others (What rules are you using?).

a. sulfur trioxide SO_3
b. diphosphorous pentaoxide P_2O_5
c. chorine trifluoride
d. disulfur difluoride

$S_2F_2 = SF$

Organizing Our Knowledge

Lewis Structures (Sections 2.2 & 2.3)

Practice 7 Warm Up

To be successful at drawing Lewis structures we need to be able to determine the number of valence electrons of an atom and an ion. You do *not* have to memorize these—simply refer to the periodic table.

Using your periodic table, fill in the following chart.

Species (atom or ion)	Number of Valence Electrons
F^- (fluoride ion)	8
O^{2-} (oxide ion)	
C (carbon atom)	
Li^+ (lithium cation)	
B (boron atom)	
Ne (neon atom)	

Species (atom or ion)	Number of Valence Electrons
F (fluorine atom)	
N^{3-} (nitride anion)	
H (hydrogen atom)	
H^- (hydride anion)	
N (nitrogen atom)	
Cl^- (chloride anion)	

Using the above chart, draw a Lewis structure for the atom or ion. Review page 62 if you need to.

Practice 8 Lewis Structures

Draw Lewis structures for the *seven* diatomic elements.

Practice 9 The Octet Rule

What is the octet rule? Which atoms *always* obey the octet rule in a molecule? Which atoms *always* obey a duet (2) rule in a molecule? Which atoms *usually* obey the octet rule, but can handle more than an octet?

Practice 10 More Lewis Structures

Consider the following Lewis structures. Every structure *has an error.* Find the error.

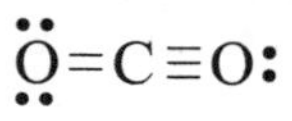

a. Ö=C≡O:

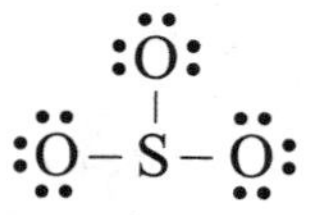

b. :Ö–S–Ö: (with :Ö: bonded above S)

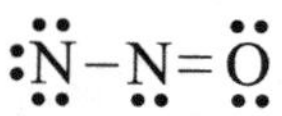

c. :N̈–N=Ö

Practice 11 Getting the Skeleton

Sometimes, the hardest part of doing Lewis structures is getting the skeleton. There are several keys you can use to deduce a skeleton:

- Usually the first atom in the chemical formula is the central atom. This is almost always true when you have a *binary molecule* and there is only one atom of the first type. See the skeleton for SiF_4 (page 78).
- You simply know the skeleton because the molecule is so common you have memorized it (water, ammonia, carbon dioxide).
- H can form only *one attachment* to an atom.
- C forms four bonds. These can be four single bonds, two double bonds, one double and two single bonds, etc. Carbon *does not* usually have lone pairs of electrons.

Sometimes the above information is enough to deduce the skeleton. Sometimes, however, it is not. When it is not, then the following must occur:

- You are told the central atom or you are given the skeleton. Ah, this makes Lewis structures much easier!

Draw the skeleton for each of the following molecules or ions.

a. SiO_2

b. H_2O

c. CH_3Br

d. CO_2

e. OF_2

Practice 12 Putting It Together and Getting the Structure

Draw Lewis structures for the following molecules and ions.

Draw Lewis structures for the molecules in Practice 10.

a. BrCN (C is the central atom)

b. $GeCl_3^-$

c. BrO_2^-

d. CHI_3 (C is the central atom)

e. CH_3NH_2

Practice 13 When More Than One Lewis Structure is Possible

For many molecules, it is possible to draw more than one *valid* Lewis structure. Let's look at the case for ozone (O_3). One possible Lewis structure is:

O::O:O

a. Verify for yourself that this is an accurate Lewis structure. Is this number of valence electrons correct? Is the octet rule satisfied?

We can describe the *bonding* in this Lewis structure as follows:
There is one double bond between the first and second oxygen,
There is one single bond between the second and third oxygen.

Note that I could just as accurately draw the Lewis structure of ozone as:

O:O::O

How would I interpret the bonding in this case?

We have a situation where there is more than one *valid* Lewis structure for a given molecule. And notice that each Lewis structure gives a *different* picture of the bonding. So what is correct? Does ozone really have a double and a single bond? And, if so, which bond is double and which is single?

The answer to this question is quite fascinating. Ozone does *not* have a double and a single bond—both bonds in ozone are identical. Thus, neither Lewis structure accurately represents the bonding in ozone. On the other hand, if we imagine that the true structure is a mixture of both Lewis structures, we get a more accurate picture. How could we mix the two Lewis structures? The first Lewis structure states that the bond between the first O and the second is a double bond. The second structure suggests that the bond is a single bond. If we *average* these two, we get something between a single and a double. If we play the same averaging game for the bond between the second O and the third oxygen, we also get a bond between a single and a double. Notice that the bond between the first and second O and the bond between the second and third oxygen are the same when we average these structures!

This averaging idea is the concept of "resonance." One way to think of resonance is that each individual Lewis structure is not a valid picture of reality but if we average all the valid Lewis structures together we then get a more accurate picture. We will call each valid Lewis structure a resonance structure.

b. Consider CO_2, which has the skeleton structure: O—C—O. There are three valid Lewis structures (three resonance structures) for this molecule. What are they?

c. Consider N_2O, which has the skeleton structure: N—N—O. There are two resonance structures for this molecule. What are they?

CHAPTER 3

Ionic Compounds

Checklist

What you need to be able to do when you finish Chapter 3

- Recognize binary ionic compounds.

  Example: Which of the following three compounds are binary ionic compounds: MgO, SO_2, $NaClO_4$?

- Using the periodic table, know the most common ions for the elements in Groups I, II, V, VI, VII.

  Example: What is the ion for each of the metals in Group I (Li,Na, K)?

- Use your knowledge of the ions to construct binary ionic compounds.

  Example: Construct the ionic compound from the ions of potassium (K) and oxygen (O).

- Name binary ionic compounds.

  Example: What is the name of the compound you constructed in Example 3?

- Know (memorize) common polyatomic ions and be able to name and recognize the ions in an ionic formula.

  Example: What is the formula for the nitrate ion? (*Be sure to include the charge—it is part of the formula of an ion.*) Which of the following two formulas contains the nitrate ion—KNO_3 or NO_3?

- Construct binary ionic compounds using both monatomic and polyatomic ions.

  Example: Construct the binary ionic compound from the magnesium cation and the sulfate anion.

Organizing Our Knowledge

Recognizing and Naming Ionic Compounds

A key goal of this chapter is recognizing an ionic compound from its chemical formula. This can get complicated, but, in general, there are only a few rules. Here are the few rules of ionic compounds:

1. The compound is binary (containing only two *types* of atoms): one type of atom is from the metal side of the periodic table and the other type of atom is a nonmetal.
2. The compound is not binary but contains a polyatomic anion and a metal cation. To see this pattern, you have to be able to recognize a set of polyatomic anions.
3. The compound is not binary but contains an ammonium cation (NH_4^+) and a nonmetal or polyatomic anion. This is one of the most complicated cases you can get.

We can arrange these rules in a flow diagram:

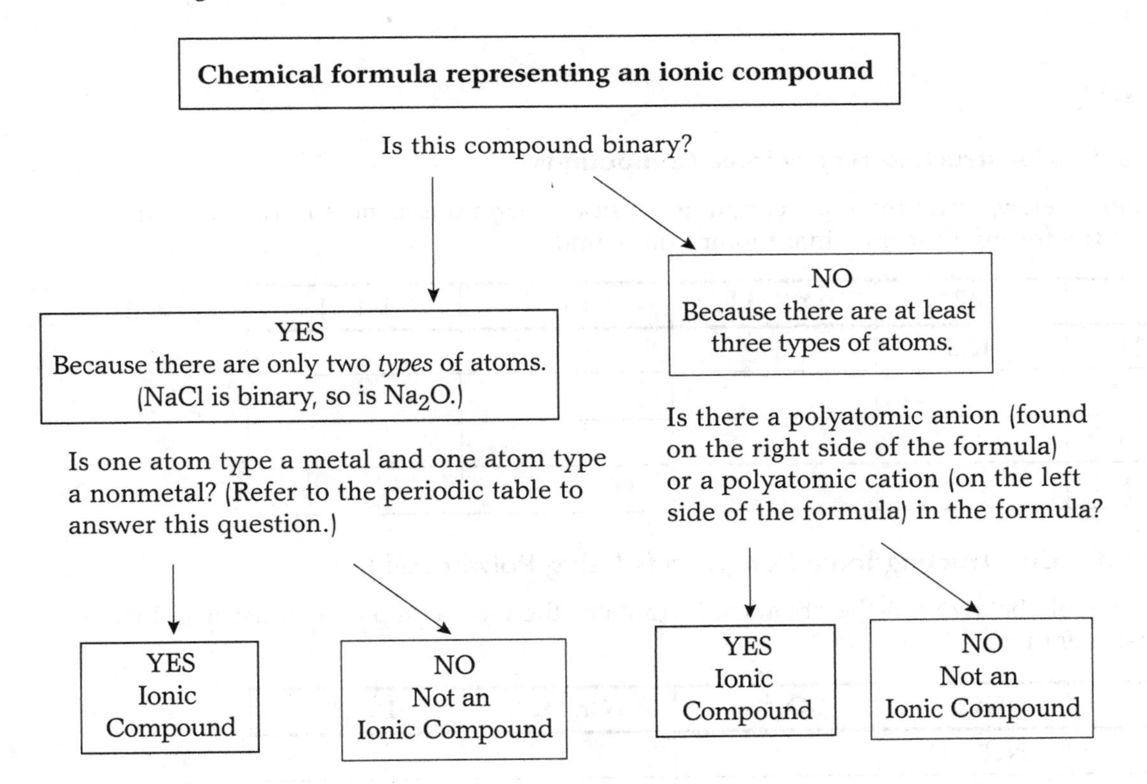

Practice Problems and Study Hints

Practice 1 Identifying Ionic Compounds

It is essential that you master the above diagram and can answer questions such as the following: Indicate whether or not the following compounds are binary ionic compounds:

a. KCl

b. SF_6

c. MgS

d. Cs_2O

e. Na_3PO_4

Find the polyatomic ion in each of the following compounds:

a. K_2SO_4

b. $Ba(OH)_2$

c. NH_4Cl

d. $Mg(NO_3)_2$

Practice 2 Constructing Binary Ionic Compounds

In the table below, write the most common ion in the parentheses next to the atom and construct the formula for the binary ionic compound.

	S (S^{2-})	O ( )	I ( )	F ( )
K (K^+)	K_2S			
Mg ( )				
Ba ( )				
Al ( )				

Practice 3 Constructing Ionic Compounds Using Polyatomic Ions

Fill in the table below with the chemical formula of the ionic compound constructed from the constituent ions.

	NO_3^-	SO_4^{2-}	$Cr_2O_7^{2-}$	PO_4^{3-}
NH_4^+	NH_4NO_3			
K^+				
Ca^{2+}				
Al^{3+}				

Practice 4 Naming Binary Ionic Compounds

The table below is repeated from Practice 2. Name each of the compounds you constructed from the constituent ions.

	S (S^{2-})	O ( )	I ( )	F ( )
K (K^+)				
Mg ( )				
Ba ( )				
Al ( )				

Practice 5 Naming Ionic Compounds that Contain Polyatomic Ions

Repeat Practice 4 with the table containing polyatomic ions.

	NO_3^-	SO_4^{2-}	$Cr_2O_7^{2-}$	PO_4^{3-}
NH_4^+				
K^+				
Ca^{2+}				
Al^{3+}				

Practice 6 Learning Names and Charges of Polyatomic Ions

Your instructor will have you memorize a set of polyatomic ions. Write the list with their names (and charges!) below.

Practice 7 Constructing Ionic Compounds from Polyatomic Ions

Construct a table, such as the one in Practice 5, using all of the polyatomic ions that you must memorize. Put all the anions across the top and the cations along the left. Include any metal cations that you need to make the table have at least four rows. Construct ionic compounds from these constituent ions. Name each of your compounds.

CHAPTER 4

The Mole and Chemical Equations

Checklist

What you need to be able to do when you finish Chapter 4

- Count the number of atoms in a molecular unit

  Example: How many carbon atoms are in one molecule of C_3H_6O?

- Count the number of ions in a formula unit

  Example: How many nitrate ions are in one formula unit of $Mg(NO_3)_2$?

- Count the number of moles of atoms in a mole of a molecular compound

  Example: How many moles of carbon are in one mole of C_3H_6O?

- Count the number of moles of ions in a mole of an ionic compound

  Example: How many moles of nitrate ions are in one mole of $Mg(NO_3)_2$?

- Count the number of units in a mole of a compound

  Example: How many carbon atoms are there in one mole of C_3H_6O?

Note: Compare your answers to the first, third, and fifth examples. All the answers should be different. Are your answers different? Read the question *carefully* to determine if you are being asked for moles or numbers of particles.

Note: Continue to review your nomenclature, particularly the polyatomic ions.

- Write ratios expressing the relative numbers or moles of atoms when you are given a molecular formula.

  Example: What is the ratio of carbon atoms to hydrogen atoms in C_3H_6O?

- Determine the chemical formula from the ratios of moles of atoms or ions.

  Example: One mole of octane contains eight moles of carbon and 18 moles of hydrogen. What is the chemical formula of octane?

- Write ratios expressing the relative numbers or moles of ions when you are given a formula unit.

  Example: What is the ratio of magnesium ions to nitrite (NO_2^-) ions in $Mg(NO_2)_2$?

- Determine the chemical formula from the ratios of moles of ions.

  Example: One mole of silver sulfate contains two moles of silver ion and one mole of sulfate ion. What is the ionic formula for silver sulfate?

- Translate a description of a chemical reaction into a chemical equation.

  Example: One mole of sulfuric acid reacts with two moles of sodium hydroxide to produce one mole of sodium sulfate and two moles of water. Write the balanced chemical equation for this reaction.

- Write the chemical equation using Lewis structures.

  Example: Write Lewis structures for each of the compounds in the following chemical reaction, and show that atoms are neither lost nor gained in the reaction: $2NH_3 + 3O_2 + 2CH_4 \rightarrow 2HCN + 6H_2O$.

- Use the coefficients in a balanced chemical equation to write ratios of reactants and/or products.

  Example: What is the ratio of moles of aluminum reacting to moles of aluminum bromide being produced in the reaction: $2Al + 3Br_2 \rightarrow 2AlBr_3$?

- Balance a chemical equation.

  Example: Balance the reaction: ___ Fe + ___ $S_8 \rightarrow$ ___ FeS.

Concept Maps

Concept maps are useful visual tools for seeing the relationships among concepts. For example, let's construct a concept map for all the information obtained in a chemical formula. Let's relate the following concepts and the example given:

Chemical formula (*e.g.*, $C_6H_{12}O_6$)
Atoms types
Number of atoms per molecule
Ratio of atoms in a molecule
Number of moles of atoms per mole of molecules
Number of molecules per mole of molecules
Ratio of moles of atoms in a mole of molecules

My concept map looks like this:

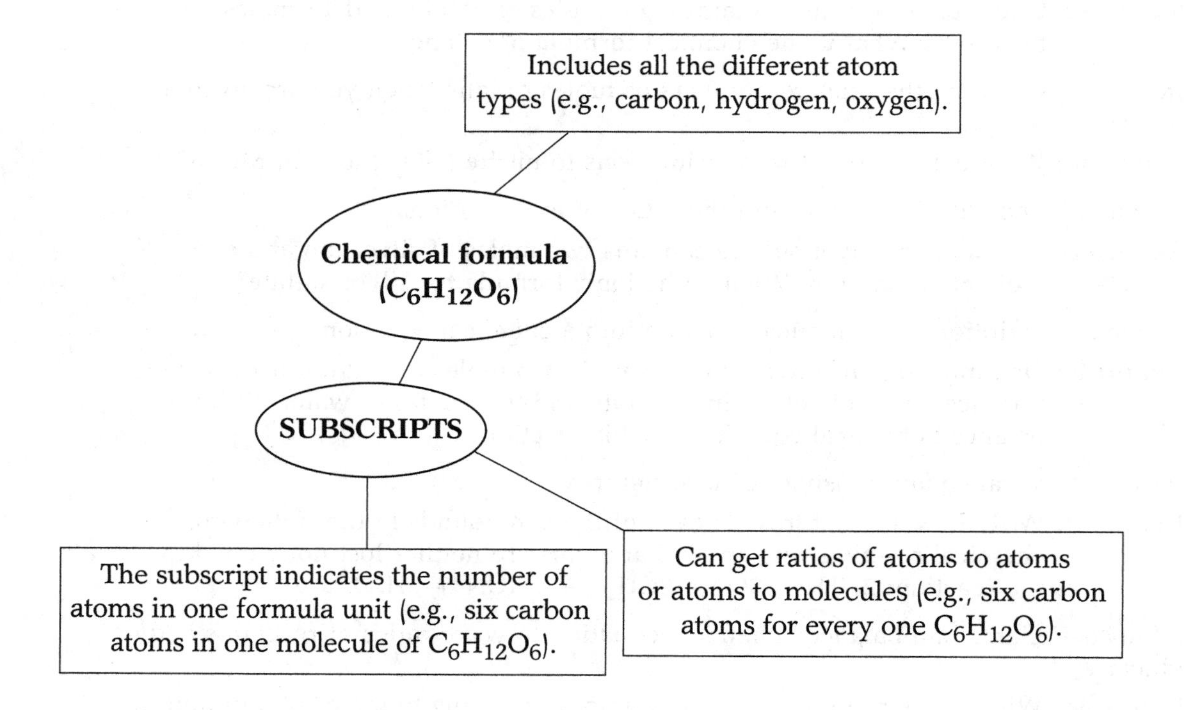

Note that there is no one way to make a concept map. Your map may look a bit different. However, it is the process of constructing the map and exploring the relationships that will help you cement your knowledge of the concepts of chemistry.

Finish constructing the concept map using all the concepts listed above.

Draw a concept map relating to the coefficients in a balanced chemical equation. Use the following concepts and example given:

Balanced chemical equation (*e.g.*, $2C_3H_8 + 10O_2 \rightarrow 6CO_2 + 8H_2O$)
Mole ratios of reactants
Mole ratios of reactants and products
Mole ratios of products
Same number of atoms on right side and left side of the chemical equation
No atoms are lost or gained in a chemical reaction

Practice Problems and Study Hints

Practice 1 Counting Atoms and Ions in Molecular and Ionic Formulas

Hint: Start to look for patterns in the questions. Keep practicing these types of questions until you begin to feel that they are repetitive. Once they have become repetitive, you have mastered the concept!

1. Imagine you have a beaker containing 150 molecules of carbon tetrachloride (CCl_4).
   a. How many carbon atoms are in the beaker?
   b. How many chlorine atoms are in the beaker?
   c. What is the ratio of carbon atoms to chlorine atoms in the beaker?
   d. What is the ratio of carbon atoms to chlorine atoms in the chemical formula?
   e. Are your answers to c & d the same? Why or why not?
   f. At room temperature, carbon tetrachloride is a gas. Do you think you could actually see 150 molecules of carbon tetrachloride?
   g. Replace carbon tetrachloride with any other molecule chosen from your textbook and answer all the questions again.
2. The chemical formula for propylene glycol is $C_3H_8O_2$.
   a. How many moles of carbon atoms are in one mole of propylene glycol?
   b. How many moles of carbon atoms are in two moles of propylene glycol?
   c. How many moles of hydrogen atoms are in two moles of propylene glycol?
   d. What is the mole ratio of carbon to hydrogen in propylene glycol?
   e. How many moles of oxygen atoms are in one mole of propylene glycol?
   f. How many moles of oxygen atoms are in 1/2 mole of propylene glycol?
   g. Replace propylene glycol with another molecule chosen from your book and answer all the questions again.
3. Look through your book and pick a molecule that has at least three different types of atoms. Ask all the possible questions about the *number of individual atoms* and about the number of *moles* of atoms in the molecule. Share your questions with your peers. Have you investigated every possible way to ask a *quantitative* question about counting the numbers and moles of atoms in a formula?
4. Repeat the above three exercises using ionic compounds. Here is an opportunity to rexview the polyatomic ions.
   First, pick an ionic compound (*e.g.*, iron (III) sulfate [$Fe_2(SO_4)_3$]) and any amount (*e.g.*, 750 formula units).
   a. How many iron (III) (Fe^{3+}) ions are there in 750 formula units of iron (III) sulfate?
   b. How many sulfate ions are there?
   c. What is the ratio of iron to sulfate ions?
   d. Does the ratio of iron to sulfate ions change if I change the number of formula units?

e. Next, consider the quantity to be three moles of the iron (III) sulfate. How many moles of iron ions are there in three moles of iron (III) sulfate? How many moles of sulfate?

f. What is the mole ratio of iron (III) to sulfate? How does your answer compare to (c)?

g. What is the mole ratio of iron (III) to one mole of iron (III) sulfate?

h. Repeat the questions but change the ionic compound, the number of formula units, and the number of moles.

Practice 2 Translating Descriptions into (Unbalanced) Chemical Equations

Hint: It is very important to be able to read a description of a chemical reaction and then write the equation using formulas. To do this you have to:

1. Identify the reactants and products.
2. Translate the name of the compound into its formula. To do this, you have to use your nomenclature.
3. There is always some confusion about translating the words "burns," "combusts," and "heat" into the chemical reaction. Burns and combusts *always* means "reacts with oxygen." Think about snuffing out a candle. The candle no longer burns because there is no oxygen available. Heat, on the other hand, does *not* mean "reacts with oxygen." Thus, "decomposes when heated" is just a description and does not imply reaction with a chemical species.

For each description below, identify reactants and products, and try to write the chemical equation. Don't worry about balancing the equation yet; save that for the next exercise. In each case, ask yourself if oxygen is a reactant.

1. Octane (C_8H_{18}) is the fuel for the internal combustion engine. Complete combustion of the octane produces carbon dioxide and water.
2. Octane when burned produces carbon dioxide and water.
3. Incomplete combustion of octane produces carbon monoxide and water.
4. The metabolism of glucose ($C_6H_{12}O_6$) in the body is similar to a combustion reaction—the combustion of glucose results in the production of carbon dioxide and water.
5. Silicon is produced by the reaction of silicon dioxide and carbon in a furnace under extremely high heat. Carbon monoxide is also produced in this reaction.
6. Large quantities of sodium hydroxide are used in the chemical industry. Sodium hydroxide is produced by adding electricity to salt water (NaCl and water) to produce chlorine gas, hydrogen gas, and the desired sodium hydroxide.
7. Potassium metal reacts vigorously with water to produce a purple flame, hydrogen gas, and potassium hydroxide.
8. Iron can be produced from iron ore (Fe_2O_3) by reacting the ore with carbon monoxide. In addition to the iron, carbon dioxide is produced and enters the atmosphere.

Practice 3 Using Lewis Structures to Write Balanced Chemical Equations

Writing balanced chemical equations using Lewis structures is an excellent way to practice your Lewis structures, see the balance of each of the atom types on the two sides of the chemical equation, and see the meaning of the coefficients in the chemical equation.

Consider the reaction of nitrogen gas (N_2) and hydrogen gas (H_2) to produce ammonia gas (NH_3).

a. First, review Lewis structures by writing the Lewis structures for the nitrogen, hydrogen, and ammonia gases.

b. Write the chemical reaction (unbalanced) using the Lewis structures.

c. Write the chemical reaction (unbalanced) using just the chemical formulas.

d. Balance the chemical equation.

e. Write the balanced chemical equation using Lewis structures.

f. Write a sentence describing the balanced chemical equation (see the bottom of page 130).

g. Write a sentence describing the meaning of the coefficients in the balanced chemical equation.

Practice 4 Balancing Equations

Look at the equations you wrote in Practice 2. Try to balance each of these equations. Use the method given in Table 4.2 (page 129).

Hint: Balancing equations has a built-in check, so you should be certain if your balanced equation is correct. Count the number of atoms of each type on both sides of the equation. If they are the same, your equation is balanced. If not, keep trying to balance the equation. Persistence and checking always works!

CHAPTER 5

Chemical Reactions

Checklist

What you need to be able to do when you finish Chapter 5

- Read a description of a change and identify it as either chemical or physical.
- Look at a chemical equation and identify the reaction as synthesis, decomposition, single displacement, double displacement, or oxidation-reduction.
- Predict products of certain reaction types.
- Assign oxidation states to elements, metals, and ions in ionic compounds.

Concept Map

Finish the following concept map relating the words used to describe chemical or physical changes.

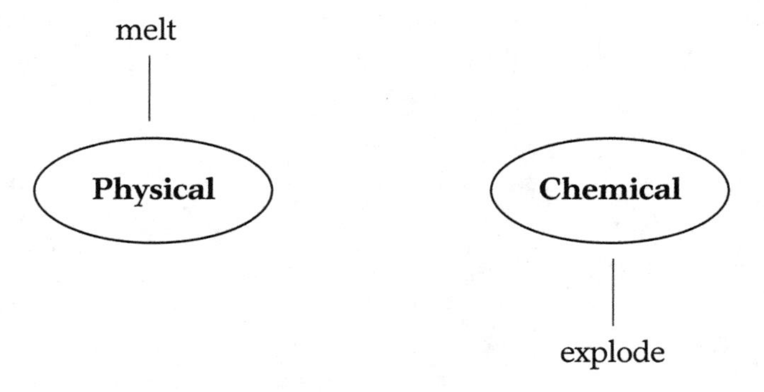

Words to include: burn, freeze, thaw, form, condense, change color, react

Practice Problems and Study Hints

Practice 1 Types of Reactions

First, notice that the six types of reactions in Table 5-2 are not independent. For example, a synthesis reaction can also be an oxidation-reduction reaction. A single displacement reaction is always an oxidation-reduction reaction. A double-displacement reaction is rarely an oxidation-reduction reaction. In fact, synthesis, decomposition, single-displacement, and double-displacement are *mutually exclusive* reaction types. However, they can also be oxidation-reduction reactions. Thus, let's first learn about the four mutually exclusive types, and then consider oxidation-reduction reactions.

We can systematically identify reaction types by using the following flow diagram:

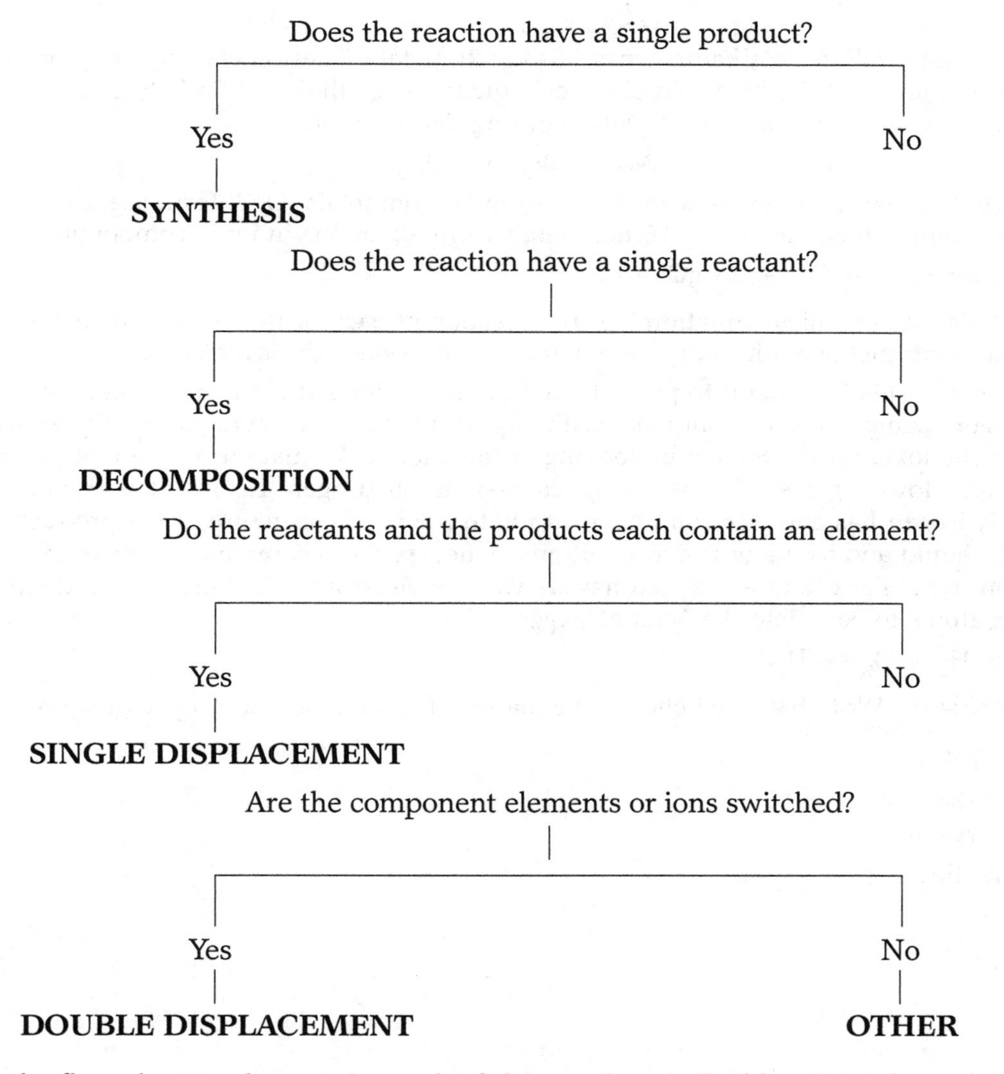
Does the reaction have a single product?
Yes
No
SYNTHESIS
Does the reaction have a single reactant?
Yes
No
DECOMPOSITION
Do the reactants and the products each contain an element?
Yes
No
SINGLE DISPLACEMENT
Are the component elements or ions switched?
Yes
No
DOUBLE DISPLACEMENT
OTHER

Use the flow chart to characterize each of the reactions in Problem 8, at the end of Chapter 5. When you have identified a reaction as a synthesis type, write the reaction in reverse as a decomposition type.

Practice 2 Predicting Products of Synthesis Reactions

1. *A metal and a nonmetal.* The trick to this type of reaction is determining the appropriate subscripts. We know that the product of Rb + Cl_2 has to be rubidium chloride. But what are the subscripts?

   In ionic compounds, all halogens (Group 17) have a charge of –1. And all Group 1 metals (such as Rb) have a charge of +1 in ionic compounds. Our resultant product has to be *electrically neutral.* When we add up all the charges, the sum must be zero (0). In the case of rubidium chloride, the resulting formula has to be RbCl.

   a. Consider all the alkali (Group 1) metals. Write balanced chemical equations for the reactions between each of these and one halogen of your choosing. Note that the elemental state for the halogens is *diatomic.* For example:

   $$2Cs + I_2 \rightarrow 2CsI$$

   b. Next, consider all of the alkaline earth (Group 2) metals. These metals form +2 ions in ionic compounds. Write balanced chemical equations for the reaction between each of these metals with one halogen of your choosing. For example:

   $$Mg + Br_2 \rightarrow MgBr_2$$

   c. If we realize that, like the halogens, the Group 16 nonmetals (O, S) form specific anions in ionic compounds, the Group 16 nonmetals form –2 anions in ionic compounds.

   Example: $2Mg + O_2 \rightarrow 2MgO$

   Write balanced chemical equations for the reaction of each of the alkali and each of the alkaline earth metals with oxygen. Remember that oxygen gas is *diatomic.*

2. *Two nonmetals.* This is difficult to predict because many times there is more than one product. For example, carbon can react with oxygen to form carbon dioxide (CO_2) as well as carbon monoxide (CO). So just by looking at the chemical equation, we cannot predict the product. However, there is one special case—when hydrogen (H_2) reacts with nonmetals. Hydrogen has one electron to contribute to the Lewis structure of the product. Thus, we should add as many hydrogen atoms as necessary to complete the octet of the other atom type. For example, oxygen has six valence electrons. We therefore need two hydrogen atoms to complete the octet of oxygen.

   Example: $H_2 + O_2 \rightarrow H_2O$

   Try It Yourself: Write balanced chemical equations for the reaction of hydrogen gas with:

   a. oxygen gas

   b. chlorine gas

   c. iodine crystals

   d. bromine liquid

Practice 3 Predicting Products of Single-Displacement Reactions

The products of these reactions are almost obvious—like displaces like. A nonmetal element will displace the nonmetal from the compound. A metal element will displace the metal ion from the compound. Observe how two reactions below follow this pattern:

$$2\ Al_2Cl_3 + 3\ Ca \rightarrow 3\ CaCl_2 + 4Al$$
$$2\ Al_2Cl_3 + 3F_2 \rightarrow 2Al_2F_3 + 3Cl_2$$

In the first reaction, the calcium metal kicks out the aluminum metal from the aluminum chloride. How do we know the chemical formula of the reactant? Calcium chloride is an ionic compound (a metal and a nonmetal). Calcium is from group 2 of the periodic table and therefore has a +2 charge. Chloride has a −1 charge. Therefore, the chemical formula (electroneutrality) is $CaCl_2$.

There is a pattern to this.

- First, predict the products (in words). For example, we know that the calcium must kick out the aluminum, therefore the product contains calcium and chloride.
- Second, use your knowledge of the charges to predict the chemical formula.
- Third, balance the chemical equation.

Try this with the second equation.

- First, the fluorine (a nonmetal) must kick out the chloride (also a nonmetal). The product must therefore contain aluminum and fluoride.
- Second, chloride has a −1 charge, therefore the aluminum must have a +3 charge in aluminum chloride. Thus, the Al must still have a +3 charge in aluminum fluoride, and fluoride (Group 17) must have a −1 charge. Hence, the formula is Al_2Cl_3. (This is easier. Because F comes from the same group as Cl, the formula has to stay the same: $Al_2Cl_3 \rightarrow Al_2F_3$.)
- Third, balance the equation.

Try It Yourself: Practice with these equations. Write the balanced chemical equation for the following single-displacement reactions. Use the three-step method (figure out the product, figure out the chemical formula, then balance the equation).

a. $Hg_2Cl_2 + Al \rightarrow$

b. $CuO + Na \rightarrow$

c. $CoCl_4 + F_2 \rightarrow$

Practice 4 Predicting Products of Double-Displacement Reactions

Precipitation reactions are also called double-displacement reactions. Notice what happens in a precipitation reaction. You mix two solutions together and a cloudy precipitate forms in the reaction vessel. There is a pattern to these kinds of reactions.

First, you are mixing two solutions that contain water-soluble ionic compounds. For example, you might have NaCl in water (common table salt in water). We know the sodium chloride dissolves in the water. When it dissolves it breaks into its ions. We, therefore, need to be able to identify the ions. Fill in the table below with the appropriate anions and cations. Review your polyatomic ions if you need to.

Ionic Compound	Cation	Anion
NaCl	Na^+	Cl^-
K_3PO_4		
$MgSO_4$		
$AgNO_3$		
NH_4OH		

Second, suppose we mix two solutions. Imagine that just before anything happens we look into the solution. What is present?

Consider the reaction of $AgNO_3$ (*aq*) + KCl (*aq*).

Just after mixing, four ions are present: Ag^+, K^+, NO_3^-, Cl^-.

If any of these ions can combine to form an insoluble salt, then a precipitate will form. We could consider all possible combinations but we really only have to consider swapping the anions. (Why is that?) So swap the nitrate with the chloride and ask if any of the new compounds are water *in*soluble.

$$AgNO_3\ (aq) + KCl\ (aq) \rightarrow AgCl + KNO_3$$

To determine if the product will precipitate we need to consider the solubility rules. Let's look up each of the products in Table 5-3—the solubility trends. The first rule that applies to silver chloride is rule 3: "The chlorides of all metals are soluble *except silver.*" Therefore, silver chloride is *insoluble* and precipitates out of the solution. We should then add an (s) after silver chloride:

$$AgNO_3\ (aq) + KCl\ (aq) \rightarrow AgCl\ (s) + KNO_3$$

What about the potassium nitrate? The first rule states that nearly all compounds of potassium are soluble. The second rule states that nitrates are soluble. Therefore, potassium nitrate is water-soluble and we should write an (*aq*) in the chemical equation:

$$AgNO_3\ (aq) + KCl\ (aq) \rightarrow AgCl\ (s) + KNO_3\ (aq)$$

Practice writing balanced chemical equations with the following double-displacement reactions. Use the three-step method: obtain the product, predict the solubility, and balance the equation.

a. $Ba(NO_3)_2\ (aq) + NaOH\ (aq) \rightarrow$

b. $CaCl_2\ (aq) + K_2CO_3\ (aq) \rightarrow$

c. $Pb(NO_3)_2\ (aq) + MgSO_4\ (aq) \rightarrow$

Practice 5 Oxidation-Reduction Reactions

In most of the examples above, the chemical reactions were also oxidation-reduction reactions. Oxidation-reduction reactions are ones in which electrons are transferred from one atom or ion to another atom or ion. Chemists keep track of this electron transfer by assigning oxidation states to the atoms in the chemical formulas. We already have some experience assigning oxidation states. Here are a few rules that we have used:

1. The oxidation states of metal ions is the charge of the ion. Therefore, the oxidation state of the Group 17 ion in compounds of chlorides, fluorides, bromides, or iodides is always -1.
2. The oxidation state of elements is always zero.
3. The oxidation state of hydrogen is +1 when hydrogen is bound to a nonmetal (HCl). When hydrogen is bound to a metal, its oxidation state is -1 (LiH).
4. The oxidation state of oxygen is usually -2.

We can use these rules to decide if a reaction is an oxidation/reduction reaction. Let's return to our synthesis reactions:

$$4Al + 3O_2 \rightarrow 2Al_2O_3$$

Let's assign oxidation states to each atom by writing the oxidation state above the symbol. Because both reactants are elements, the oxidation states must be zero:

$$\overset{0}{4Al} + \overset{0}{3O_2} \rightarrow 2Al_2O_3$$

Next, let's assign oxidation states to the aluminum oxide. We know the oxidation state of oxygen (rule 4).

Oxidation state of O:	−2
Number of O	3
Total negative charge	−6

Therefore:

Total positive charge	+6
Number of Al	2
Oxidation state of Al	+3

This reasoning is identical to constructing the chemical formulas of ionic compounds.

Rewrite the chemical equation to include the oxidation states of all atoms in all compounds:

$$\overset{0}{4Al} + \overset{0}{3O_2} \rightarrow 2\overset{+3}{Al_2}\overset{-2}{O_3}$$

We are now ready to ask if this is an oxidation-reduction reaction. If it is, then the oxidation state of the atoms must change. In fact, one must increase and one must decrease.

Consider aluminum (Al). Its oxidation state changed: $0 \rightarrow 3$.

Similarly, O: $0 \rightarrow -2$.

We conclude that this is an oxidation-reduction reaction.

In fact, any reaction in which an element reacts with something to form a compound or in which an element is produced from a compound has to be an oxidation-reduction reaction. (Why?)

Notice that one atom's oxidation state increased and one decreased.

The oxidation state of Al increased from 0 to 3.
The oxidation state of O decreased from 0 to –2.

Al is said to be *oxidized* (lost electrons).
O is said to be *reduced* (gain electrons).

Consider the following single-displacement reaction:

$$3Na + Fe_2Cl_3 \rightarrow 2Fe + 3NaCl$$

a. Is this an oxidation-reduction reaction?

b. What are the oxidation states of all the atoms on both sides of the equation?

c. What species was oxidized?

d. What species was reduced?

For more practice, go back to your synthesis reactions and single-displacement reactions. All of these are oxidation-reduction reactions. Determine the species oxidized and the species reduced in each reaction.

CHAPTER 6

Quantitative Properties of Matter

Checklist

What you need to be able to do when you finish Chapter 6

- Use the density equation to compute mass or volume.

  Example: The density of aluminum at 20 °C is 2.70 g/cm^3. How many grams of aluminum are there in 5.2 cm^3? Or, how many cm^3 do you need to obtain 10.4 grams of aluminum?

- Compute the density of a material from the measurement of volume displaced.

  Example: When 22.24 grams of manganese is immersed in a graduated cylinder containing 21.20 cm^3 of water, the water level rises to 24.18 cm^3. What is the density of manganese?

- Interpret the chemical formula in terms of ratios of atoms to molecules (or atoms to atoms).

  Example: What is the ratio of nitrogen atoms to nitrogen dioxide molecules (NO_2)? What is the ratio of nitrogen atoms to oxygen atoms in nitrogen dioxide?

- Use the chemical formula to compute numbers of atoms in a large collection of molecules.

  Example: How many hydrogen atoms are there in 8534 molecules of C_3H_8O?

- Use the chemical formula to compute the number of molecules needed to generate a specific number of atoms.

  Example: How many molecules of C_3H_8O do you need to have 76,496 atoms of hydrogen?

- Use the chemical formula to compute the number of dozens or number of moles of atoms in a given number of dozens or moles of molecules.

  Example: How many dozen carbon atoms are there in six dozen C_3H_8O molecules? (Try this: How many carbon atoms are there in 12.4 dozen C_3H_8O molecules?)

This chapter is primarily concerned with learning to use *ratios and proportions* to compute chemical and physical quantities. ***This kind of proportional reasoning is at the core of quantitative chemistry.*** Therefore, you are going to want to practice this kind of reasoning throughout the entire course. However, before you practice the quantitative calculations, it is helpful to know about the types of variables discussed in chemistry. It is also important to have a general feel for the *physical state* (liquid, solid, gas) of a material at a given temperature.

Concept Map

Draw a concept map relating the common kinds of variables used and measured in chemistry. Use the following terms/concepts in your map:

Chemical variables
Extensive variables
Intensive variables
Temperature
Mass
Density
Volume
Grams
cm^3
Number of molecules
Ratio of atoms per molecule in a chemical formula

Relating Temperature to Physical State

Look at the periodic table and indicate which elements are liquids, which are solids, and which are gases at room temperature. Pick three elements: one that is a liquid, one that is a solid, and one that is a gas at room temperature. Make a table of these three elements like so:

Element	Melting Point	Boiling Point

For each of your elements indicate whether the melting point and boiling point are greater than or less than room temperature.

Practice Problems and Study Hints

Practice 1 Working with Density

Hint: Only a few types of problems can be asked about density. These types of problems are:

1. Compute density of a substance.
2. Given the density of a substance, compute mass or volume.

Notice that in both cases it is a matter of manipulating the equation:

$$d = m/V$$

It is important to memorize the density equation. Moreover, it is important to rearrange the density equation *every time you need to do so* to solve the problem. For example:

How many grams are there in 22.5 cm^3 of sulfur? Sulfur has a density of 1.96 g/cm^3.

To solve this problem, let's start with the density equation:

$$d = m/V$$

Rearrange it to get what we want:

$$m = d \times V$$

And then plug in the numbers and solve:

$$m = 1.96\ g/cm^3 \times 22.5\ cm^3 = 44.1\ grams$$

Notice that we always start with the density equation. It is tempting to try to memorize all the permutations of the equation. For example:

$$d = m/V$$

$$m = d \times V$$

$$V = m/d$$

But *don't* do this. Practice rearranging every time—then your algebra will just get better and better! Besides, soon there will be too many equations to memorize if we have to remember every permutation.

Now practice the following problems setting up the problem from the density equation every time.

1. A piece of solid matter is found to weigh 2.3 grams. The solid is immersed in a graduated cylinder containing water. The water level rises 1.45 cm^3. What is the density of the solid?
2. The same solid (for which you now know the density) is put into a graduated cylinder containing water. Originally, the volume of the water was 15.80 cm^3. What is the reading of the water level after the solid is added?
3. Let's keep working with the same solid. Suppose 12.2 grams of the solid are put into a graduated cylinder containing 12.25 cm^3 of water. How high will the water level rise? What will be the reading of the water level in the cylinder?

4. Suppose I have 5.6 cm^3 of the solid. How much will it weigh?
5. Make up your own problems working with the same solid. In particular, consider the following types of scenarios:
   a. Design the problem to calculate the mass of the solid. Solve the problem.
   b. Design the problem to calculate the volume of the solid. Solve the problem.
   c. Design the problem to compute the density of the solid from the volume of water displaced and the mass of the solid. Solve the problem.
6. Look at the problems at the end of Section 6.2. Do not solve the problem—just look at the problem and decide which of the following it is:
   a. Compute density from mass and volume.
   b. Compute mass from density and volume.
   c. Compute volume from density and mass.
   d. other

Practice 2 Recognizing, Setting Up, and Using Ratios

The key to solving many quantitative problems in chemistry and in other sciences is using ratios (proportional reasoning). However, before you can use a ratio, you have to be able to recognize that a ratio is embedded in the nature of the problem. For example, your chapter points out two key words that are immediate clues to a ratio: *per* and *percent.* Other clues are the units ppt or ppm or ppb. Note that these units all contain the word "per" and are therefore ratios.

The word "percent" sometimes causes problems. However, percent can be thought of as a "per." For example, 3% is 3 per 100. The key is to attach the units to each of the numbers. If, for example, you have a solid rock that is 5.45% iron by weight, then you can think of the percent as 5.45 grams of iron per 100 grams of rock. The appropriate ratio would be:

$$\frac{5.45 \text{ g of Fe}}{100 \text{ g of rock}}$$

Once we have the ratio, we can then use it. For example, suppose the above rock weighed 20.3 grams. How many grams of iron are in this rock?

From the problem, we need to identify the unknown and the known. In this case, the unknown is the mass of iron in the rock and the known is the mass of the rock. Therefore, our equation is:

$$\frac{5.45 \text{ g of Fe}}{100 \text{ g of rock}} = \frac{x \text{ g of Fe}}{20.3 \text{ g of rock}}$$

Next, we can cross multiply and solve for x:

$$5.45 \times 20.3 = 100x$$

$$x = 111/100$$

$$x = 1.11 \text{ grams}$$

Let's review the steps:

1. From the problem, obtain the appropriate ratio.
2. Identify the known and unknown in the problem.
3. Set up the ratio equation.
4. Solve for x.

Worked Example:

Practice Step 1

From the problem, obtain the appropriate ratio.

Let's take a typical problem and obtain the appropriate ratio from the problem. Here is an example: For the first exam in chemistry, 90% of the students enrolled in the course took the exam. If 81 students took the exam, how many students are enrolled in the course?

Don't jump to solve the problem. Instead write down the appropriate ratio. To do the ratio, we have to immediately realize that the "%" means ratio. But what is the ratio? How about 90 students took the exam per 100 students enrolled in the course? Or, in terms of a mathematical expression, the ratio is:

$$\frac{90 \text{ students took exam}}{100 \text{ students enrolled}}$$

Now you will practice Step 1. Look at problems 28 to 34 at the end of Section 6.3. In each problem, write the appropriate ratio. *Do not solve the problem—just write down the ratio.*

Practice Step 2

Identify the known and unknown in the problem.

Consider our sample problem:

For the first exam in chemistry, 90% of the students enrolled in the course took the exam. If 81 students took the exam, how many students are enrolled in the course?

Let's identify the known and unknown: *Known:* 81 students took the exam
Unknown: number of students enrolled

Hint: Notice that you do not need Step 1 to do Step 2. Sometimes it is easier to do Step 2 than Step 1. If you are having trouble with Step 1, then do Step 2—it might help you figure out the appropriate ratio.

Now you will practice Step 2. Again, look at problems 28 to 34 at the end of Section 6.3. Identify the unknown and known.

Practice Step 3

Set up the ratio equation.

Note that you have done all the hard work. Setting up the equation should be easy—don't make it hard. All you have to do is make sure the numerators have the same units on both sides of the equal signs and that the denominators have the same units. Again, our sample problem:

For the first exam in chemistry, 90% of the students enrolled in the course took the exam. If 81 students took the exam, how many students are enrolled in the course?

Known: 81 students took the exam
Unknown: number of students enrolled

Ratio: $\frac{90 \text{ students took exam}}{100 \text{ students enrolled}}$

Thus, the numerator is the number of students who took exam, and the denominator is the number of students enrolled. The ratio equation is then:

$$\frac{\text{90 students took exam}}{\text{100 students enrolled}} = \frac{\text{81 students took exam}}{\text{x students enrolled}}$$

Next you will practice Step 3. Again, look at problems 28 to 34. Set up the ratio equation for each problem.

Practice Step 4

Solve for x.

Cross-multiply to solve the equation for x. This step does not seem to cause too many problems. In our sample problem, after cross-multiplying we obtain the following equation:

$90x = 100 \times 81$

$90x = 8100$

$x = 8100/90$

$x = 90$

Since x means the number of students enrolled, the answer must be 90 students enrolled.

Try It Yourself: Now you complete (solve) the problems 28 to 34.

Practice 3 Writing Ratios from Chemical Formulas

Chemical ratios are no more difficult than the ratios you just completed. All we need to know how to do is interpret the chemical formula. There are two ways to interpret the formula—in terms of individual atoms or in terms of moles.

Consider the chemical formula for a complex chemical called cobalt EDTA. This has the formula $Co_4C_{10}N_2H_{12}O_8$. We can use this formula to practice chemical ratios:

a. What is the ratio of cobalt atoms to one formula unit of the cobalt EDTA? The chemical formula tells us that there are 4 cobalt atoms per each formula unit. Therefore the ratio would be:

$$\frac{4\ Co}{1\ Co_4C_{10}N_2H_{12}O_8}$$

b. What is the ratio of *moles* of nitrogen to *moles* of carbon in one mole of cobalt EDTA?

This works exactly the same way, as you saw in part (a) of Practice 3. We can read the chemical formula as one molecule or unit *or* as one mole of molecules or one mole of formula units. From the formula we read that there are 2 moles of nitrogen to every 10 moles of carbon. Therefore, the ratio is:

$$\frac{\text{2 moles of N}}{\text{10 moles of C}}$$

c. Now it's time for you to practice. Set up the appropriate ratios for each of the following using the cobalt EDTA example.
   i. The ratio of oxygen atoms to cobalt atoms
   ii. The ratio of moles of oxygen to moles of carbon
   iii. The ratio of hydrogen atoms to one formula unit of cobalt EDTA
   iv. The ratio of moles of cobalt to one mole of cobalt EDTA

Sometimes the chemical formulas are a little more complicated than the cobalt EDTA because there are parentheses. We have to interpret the parentheses correctly. Cobalt makes a bright blue compound when it bonds with ammonia and sulfate to form tetraammine cobalt (IV) sulfate. The chemical formula is $Co(NH_3)_4SO_4$.

How many hydrogen atoms are there for each molecule of $Co(NH_3)_4SO_4$?

This formula tells us that there are four ammonia molecules for each formula unit of $Co(NH_3)_4SO_4$. Therefore, the ratio of hydrogen to $Co(NH_3)_4SO_4$ is:

$$\frac{12 \text{ H atoms}}{1 \text{ formula unit of } Co(NH_3)_4SO_4}$$

Practice writing the ratios using the formula $Co(NH_3)_4SO_4$.

a. What is the ratio of nitrogen atoms to sulfur atoms?

b. What is the ratio of moles of nitrogen per one mole of $Co(NH_3)_4SO_4$?

c. What is the ratio of oxygen atoms to hydrogen atoms?

Practice 4 Using Chemical Ratios to Answer Quantitative Chemical Questions

The chemical formula tells us the ratio of atoms (or moles) in one formula unit. Knowing this ratio, we can compute the total number of atoms when we have more than one molecule.

Worked Example: Consider the sucrose molecule $C_{12}H_{24}O_{12}$.

Suppose we have 15,298 sucrose molecules. How many carbon atoms do we have?
Use the procedure we developed earlier.

Step 1: Write the ratio.

Step 2: Identify the known and unknown.

Step 3: Set up the equation.

Step 4: Solve for x.

- ***Step 1:***

  We need the ratio of carbon atoms to one sucrose molecule:

$$\frac{12 \text{ C}}{1\ C_{12}H_{24}O_{12}}$$

- *Step 2:*

  Known: 15,298 sucrose molecules
  Unknown: number of carbon atoms

- *Step 3:*

$$\frac{12\ C}{1\ C_{12}H_{24}O_{12}} = \frac{x\ C}{15{,}298\ C_{12}H_{24}O_{12}}$$

- *Step 4:*

  $12 \times 15{,}298 = x$

  $x = 183{,}576$ carbon atoms

Worked Example: Let's try a challenging example. Consider the molecule pentaerythritol, which has the chemical formula $C(CH_2OH)_4$. What makes this challenging is that we have not encountered molecular formulas written this way before. However, let's stick to our rules—we know how to read chemical formulas. Using this compound, let's calculate how many moles of carbon atoms there are in 0.035 moles of pentaerythritol.

Known: 0.035 moles of $C(CH_2OH)_4$
Unknown: x moles of carbon atoms

Ratio:

$$\frac{5 \text{ moles of C}}{1 \text{ mole of } C(CH_2OH)_4}$$

(*Note:* We have five moles of carbon—four moles come from the atoms in the parentheses and an additional mole comes from the first carbon.)

Now that we have the ratio of carbon present in the molecule, let's set up the equation:

$$\frac{5 \text{ moles of C}}{1 \text{ mole of } C(CH_2OH)_4} = \frac{x \text{ moles of C}}{0.035 \text{ mole of } C(CH_2OH)_4}$$

And then solve:

$5 \times 0.035 = x$

$x = 0.175$ moles of C

d. Now try these.

i. How many moles of hydrogen atoms are in 1.53 moles of $C(CH_2OH)_4$?

ii. How many oxygen atoms are in 6499 molecules of $C(CH_2OH)_4$?

iii. How many molecules of $C(CH_2OH)_4$ are in a sample containing 33,916 carbon atoms?

e. Try all the problems at the end of Section 6.4.

CHAPTER 7

Counting and Measurement in Chemical Experiments

Checklist

What you need to be able to do when you finish Chapter 7

- Determine the number of significant figures in a number.

  Example: How many significant figures are there in 0.0023?
- Report the correct number of significant figures in a number following arithmetic manipulation.

  Example: Which of the following numbers is the correct answer to 9.030×1.2 (assume all numbers are measured numbers): 11, 10.8, 10.0, 10.84, or 10.836?
- Know the rules of exponents; be able to simplify expressions.

  Example: Simplify $4x^3/8x^2$.
- Report numbers using scientific notation.

  Example: Write the following numbers in scientific notation: 34.82, 0.692, 0.000043, −5246.84.
- Know the base metric units for mass, volume, and length.
- Be able to use metric prefixes by referring to a table.

  Example: Using Table 7.7, rewrite the following expressions using an appropriate metric prefix: 2.4×10^{-6} meters, 4.8×10^{9} grams.
- Use the exponents to *estimate* the numerical result of an expression.

  Example: Estimate the value of $\dfrac{(8.67 \times 10^{-4})(2.1 \times 10^{3})}{(0.25 \times 10^{2})(1.8 \times 10^{-6})}$.
- Convert from one metric unit to another using both the proportional reasoning method and the conversion factor method.
- Know Avogadro's number and be able to use it in converting between moles and numbers of atoms, molecules, ions, or any other type of particle.

Basic arithmetic manipulation is the subject of this chapter. There are very few new concepts in this chapter. However, the manipulations you need to be able to do are essential for success in chemistry. Probably the best way (and perhaps the only way) to get good at this is to practice and practice. So let's practice with several problems.

Practice Problems and Study Hints

Practice 1 Counting Significant Figures

You need to know the rules given in Tables 7-1 and 7-2. Look at these rules and then do problem 9 on page 236. If you want more practice, try this:

Write a number that has two significant figures but contains three zeroes. Because these zeroes cannot all be significant, what do the nonsignificant zeroes do?

Do you want more practice? How many significant figures do each of the following numbers have?

a. 3.0024

b. 243.00

c. 0.00432

Practice 2 Adding, Subtracting, Multiplying, and Dividing with Measured Numbers

The general rule for reporting the correct number of significant figures in a complex arithmetic equation is do all the calculations on the calculator and then round off in the end. To calculate the number of significant figures your answer should have, do the additions and subtractions first and determine the number of significant figures that result from these operations. Then do the multiplication and division.

First, practice addition and subtraction. Write the answer to the following using the correct number of significant figures:

a. $2.034 + 0.99 - 1.2358$

b. $0.0046 + 1.22 + 0.008$

c. $-2.988 + 0.040 + 332.044$

Next, practice multiplication and division:

d. $2.044 \times 0.99 / 1.2358$

e. $0.0046 /(1.22 \times 0.008)$

f. $(-2.988/0.040) \times 332.044$

Compare the significant figures in your answers to a–c to your answers to d–f. Are the resulting significant figures the same when the same numbers are used? (Look at your answers to [a] and [d] to answer this.) Articulate the differences between the rules for addition, subtraction, multiplication, and division.

Put it all together and compute the answer to the following using the correct number of significant figures:

g. $\dfrac{(0.013 + 21.1) \times 0.092}{33.01 \times 1.0042}$

h. $\dfrac{(1.42 - 942.114) + 0.08}{43.01 - 11.0042}$

i. $\dfrac{(626.1 - 10.005)(2.4 + 54.34)}{77.31 \times 8.8}$

Note: Significant figures are a consequence of precision in *measurement*. It is often confusing to students when numbers are measured numbers and when they are absolute numbers. After all, in the classroom you are not measuring anything. While you are not measuring anything in the classroom, almost *all* of the numbers you will use are considered to be measured numbers. Even Avogadro's number, which you use in this chapter, is a measured number. In chemistry, and in most sciences, the numbers we discuss are a consequence of someone making a measurement. Obvious exceptions to this are countable items (integers). I can count 12 eggs in a dozen, for example. There is no uncertainty in the number 12. If it is unclear from the context, assume the number is a measured number.

Practice 3 Rules of Exponents

Practice, practice, practice. That is the only way to be sure you can use the rules of exponents. Often there is confusion about how to apply the rules. Keep this simple idea in mind: If you are adding (or subtracting) monomials or numbers with exponents, you usually cannot do anything to simplify. If, however, you are *multiplying* or *dividing,* then the rules of exponents apply.

Work problems 19 through 24 referring to Table 7-4 if you need to. Eventually, you are going to want to be able to simplify expressions without using Table 7-4. Simplify the following expressions *without* the table. Practice your significant figures at the same time.

a. $(4.3x^{-3}y^{2})(0.988x^{4}y^{-6})$

b. $(98.76RT^{-1})(0.22R^{2}T)$

c. $(5.3xy - 0.81xy)/(66.22x^{2}y^{-2})$

d. $(42.6c^{2})^{2}$

e. $(0.62t^{-2})^{2} + \dfrac{1}{9.1t^{4}}$

Practice 4 Scientific Notation

Manipulating numbers written in scientific notation makes use of the rules of exponents. Before manipulating the equation, write the following numbers in scientific notation:

a. 42.009

b. 0.000000048

c. −684980000.

d. −0.0004400

Fill in the table below by *multiplying* the numbers together. Be sure your answer has the correct number of significant figures. Write your answer in scientific notation. Now is a good time to be sure you know how to use your calculator's exponent button.

	8.06×10^{-2}	9.004×10^{6}	1.124×10^{-5}	6.022×10^{23}
4.5	**3.6×10^{-1}**			
6.011×10^{-3}				
7.1×10^{5}				
1.66×10^{-24}				

Now fill in the same table but *divide* the column header by the row header. Again, be sure your answer has the correct number of significant figures. Write your answer in scientific notation.

	8.06×10^{-2}	9.004×10^{6}	1.124×10^{-5}	6.022×10^{23}
4.5	**1.8×10^{-2}**			
6.011×10^{-3}				
7.1×10^{5}				
1.66×10^{-24}				

Last, just to be certain you really know what you are doing, *add* the numbers together.

	8.06×10^{-2}	9.004×10^{6}	1.124×10^{-5}	6.022×10^{23}
4.5	**4.6**			
6.011×10^{-3}				
7.1×10^{5}				
1.66×10^{-24}				

Practice 5 Metric System

While most of the world uses the metric system, the United States does not. However, all scientists use the metric system, so the study of science in the United States requires students to learn the metric system. You might be surprised to discover that you already know some metric units.

1. Write down all metric prefixes that you already know from your personal experiences. Below are some possibilities to consider. In each case, what is the metric prefix? Using scientific notation, convert your value to an expression using just the base unit.

Try It Yourself: My computer hard drive is 10 GB, where GB stands for gigabytes. Thus, my computer hard drive is 1.0×10^{10} bytes.

a. How big is your computer's hard drive?
b. How much RAM do you have?
c. What is your favorite radio station? What do the call numbers mean?
d. A shot or a vaccine delivers what volume of fluid (approximately)? What does "cc" mean?

2. Rewrite the following expressions using an appropriate metric prefix. Refer to table 7-7 as you do this exercise:
   a. 6,300,000 seconds
   b. 2.5×10^{-4} liters (express this in milliters and microliters)
   c. 0.0000000044 meters (express this in nanometers and in picometers)
   d. 77.8×10^{6} bytes
   e. 80,000 grams (express this in kilograms)

   Express the following in the base units:
   a. 7.2 Gm (Gigameters) = how many meters?
   b. 92 ps (picoseconds) = how many seconds?
   c. 102 cm = how many meters?
   d. 432 µg = how many grams?

Practice 6 Conversions!

The early practice exercises are just warm-ups. You should think of the earlier exercises as analogous to learning where the notes are on a piano or learning how to shoot a basketball. You are not yet ready to play a concerto or a basketball game, but you are beginning to learn the basics. Now we need to practice a skill that is used in every aspect of chemistry—converting from one unit to another.

You actually did some conversion above in Practice 5, number 2. You might have done this intuitively—with no system for converting. To do more complex conversions, we need a system. There are two primary methods used for converting units: the proportional reasoning method and the conversion factor method.

Why use two methods for conversion? The proportional reasoning method gives you a greater understanding as to *why* conversions work. There is also evidence that students who are well versed in the proportional reasoning method do better in chemistry. On the other hand, the conversion factor method is easier and works better with more complicated conversion problems. Therefore, it is best if you master both methods.

Another idea to consider is that every time you change one unit to another you are doing a conversion. The process of converting is the same—you just need to see that the problem is a conversion problem. Listed below are some obvious conversions:

1. Conversions between a base unit and a metric prefix:

   meters ↔ centimeters
   gigameters ↔ meters
   kilograms ↔ grams

2. Conversions between two systems of units:

   meters ↔ feet

   kilograms ↔ pounds

3. Conversions between moles and number of chemical units:

   number of atoms ↔ moles of atoms

   number of molecules ↔ moles of molecules

While each of the above "types" of conversions may look different, they all require exactly the same process to achieve the conversion. In each case, you need the **conversion factor.**

Worked Example: Converting between a base unit and a metric prefix: Convert 44.6×10^3 kg to grams.

First, identify the conversion factor: 1 kg = 10^3 g.

Second, pick a method and set up the equation.

Proportional reasoning method:

Write the conversion factor as a proportion and set up an equation for your conversion:

$$\frac{1 \text{ kg}}{10^3 \text{ g}} = \frac{44.6 \times 10^3 \text{ kg}}{x \text{ g}}$$

Third, solve the equation. Be sure to write your answer with the correct units!

$x = (44.6 \times 10^3 \text{ g}) \times 10^4 \text{ g}$

$x = 44.6 \times 10^6 \text{ g}$

$x = 4.46 \times 10^7 \text{ g}$

OR, **Conversion factor method:**

Multiply the number you are converting by the appropriate conversion factor:

$$(44.6 \times 10^3 \text{ kg}) \times \frac{10^3 \text{ kg}}{1 \text{ kg}}$$

And then solve:

$$(44.6 \times 10^3 \text{ kg}) \times \frac{10^3 \text{ kg}}{1 \text{ kg}} = 44.6 \times 10^6 \text{ g} = 4.46 \times 10^7 \text{ g}$$

4. Practice converting between a base unit and a metric prefix using both the proportional reasoning and conversion factor methods. Before you begin the conversion, first write down the conversion factor.

   a. 23.43 GB = ? bytes

   b. 4506 km = ? meters

   c. 953,000 g = ? kg

   d. 256 μm = ? m

   e. 763,000 ms = ? s (s = seconds)

5. Practice converting among units with a metric prefix. Notice that now you have to do two conversions. Thus, it is easier to use the conversion factor method. But you still can do the proportional method if you want. Again, write down *all* the conversion factors first.
   a. 63 Mbytes = ? Kbytes
   b. 483 nm = ? pm
   c. 0.335 kg = ? mg
   d. 5×10^3 cm = ? μm
   e. 6.0×10^{-5} Gs = ? ps

6. Practice converting between moles and chemical units (atoms, molecules, ions). Remember that there are 6.02×10^{23} chemical units in each mole. Again, write down the conversion factor before performing the conversion.

   Worked Example: How many sodium ions are 0.52 moles of sodium ions?

   Conversion factor: 6.02×10^{23} sodium ions = 1 mole of sodium ion

   Conversion factor method:

$$0.52 \text{ moles of sodium ions} \times \frac{6.02 \times 10^{23} \text{ sodium ions}}{1 \text{ mole}} = 3 \times 10^{23} \text{ sodium ions.}$$

Try It Yourself:

a. How many molecules of carbon monoxide are in 3.2×10^{-3} moles of CO?
b. How many moles of carbon monoxide are 5.32×10^{24} molecules of CO?
c. How many ions of sodium are there in 1.24 moles of NaCl? How many ions of chlorine?
d. How many ions of calcium are there in 1.24 moles of $CaCl_2$? How many ions of chlorine? How does your answer here differ from your answer in c? Why?

Practice 7 Putting It All Together

In the following problems, pay attention to the number of significant figures and to your unit conversions. Also pay careful attention to the chemical formula and what the subscripts in the formula mean.

a. Suppose you have 6.23 *micro*moles of ammonia (NH_3)? How many molecules of ammonia is this? How many atoms of nitrogen? How many atoms of hydrogen?
b. 8.43×10^{21} molecules of C_2H_4O corresponds to how many moles of C_2H_4O? How many moles of carbon? How many moles of hydrogen?
c. How many molecules of H_2O are there in 58 *k*moles of H_2O? How many hydrogen atoms? How many oxygen atoms?

CHAPTER 8

Measurement of Chemical Substances

Checklist

What you need to be able to do when you finish Chapter 8

- Calculate the molar mass from the chemical formula.
- Include the correct number of significant figures in the calculation of the molar mass.
- Convert between grams and moles using molar mass.
- This goes in both directions—need to be able to convert moles to grams as well as grams to moles.
- Get more out of the chemical formula—calculating mass and moles of the individual elements in a compound.
- Use density to calculate mass.
- Obtain the percent composition of a compound from the chemical formula.
- Find empirical formulas from mass, from moles, and from numbers of atoms.
- Convert empirical formula to a molecular formula.

Concept Map

This chapter is the study of the *quantitative* data that is implicit in the chemical formula. The following diagram may help organize all the kinds of quantitative information relating to the chemical formula.

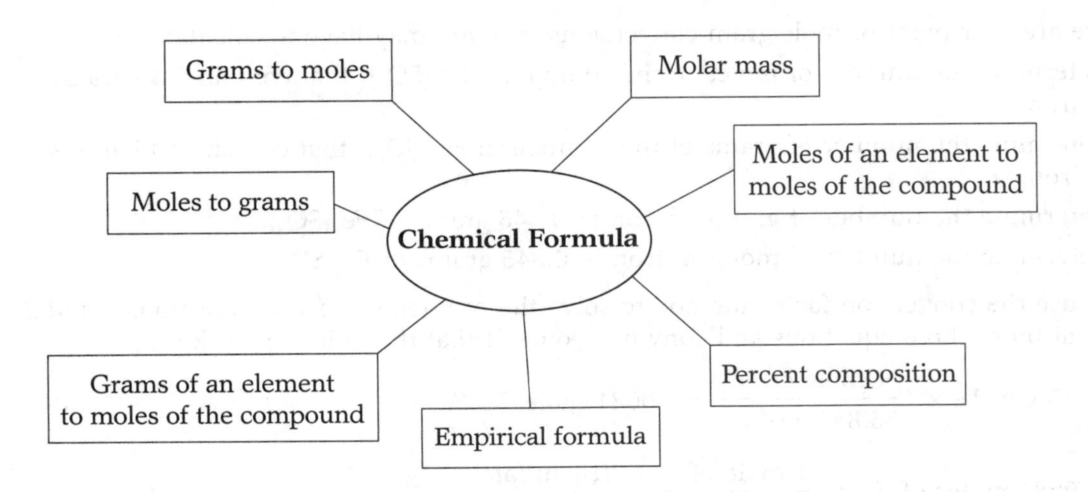

Look at this diagram and identify the kinds of calculations with which you are already quite comfortable. Which ones are you less sure about?

Practice Problems and Study Hints

Practice 1 Calculating Molar Mass

Fill in the molar mass for the vitamins on page 273. It doesn't get any more complicated than this. Pay attention to the number of significant figures of your molar mass.

Practice 2 Converting Between Grams and Moles of a Compound

The gram to mole conversion is a single-step calculation. Because grams and moles are directly proportional, you can do the conversions using proportional reasoning or the conversion factor method. You need to be good at both, so be sure to practice both methods—especially the one with which you are the least comfortable.

Do all the problems on page 274 until you feel like an expert on the gram-to-mole conversions. You need to be able to convert grams to moles and moles to grams.

Work some of the problems using proportional reasoning and some of the problems using the conversion factor method. Work as many problems as it takes until you feel like an expert on these calculations. Pay particular attention to the role of the *molar mass* in each of the calculations.

Practice 3 Relating Moles or Grams of Elements to the Moles or Grams of the Compound

These conversions always cause difficulties for students. This is a key step in the building of chemical skills. In practicing the problems, pay particular attention to the role of the molar mass in these calculations.

There are four types of mole/gram conversions that you may have to calculate:

1. Determine the number of moles of the compound $Fe_2(SO_4)_3$ that contain 0.45 grams of iron.
2. Determine the number of grams of the compound $Fe_2(SO_4)_3$ that contain 0.45 grams of iron.
3. Determine the number of grams of iron in 0.845 grams of $Fe_2(SO_4)_3$.
4. Determine the number of moles of iron in 0.845 grams of $Fe_2(SO_4)_3$.

Let's use the conversion factor method to solve the problems posed in questions 1 and 2. Look at these three equations and convince yourself that the units do work out.

$$0.45 \text{ g of Fe} \times \frac{1 \text{ mole of Fe}}{55.847 \text{ g of Fe}} = 0.0081 \text{ moles of Fe}$$

$$0.0081 \text{ moles of Fe} \times \frac{1 \text{ mole of iron (III) sulfate}}{2 \text{ moles of Fe}} = 0.0041 \text{ moles of iron (III) sulfate}$$

$$0.0041 \text{ moles of } Fe_2\,(SO_4)_3 \times \frac{399.880 \text{ g of } Fe_2\,(SO_4)_3}{1 \text{ mole of } Fe_2\,(SO_4)_3} = 1.6 \text{ g of } Fe_2\,(SO_4)_3$$

Next, look at these equations again and notice where the molar mass of the iron (III) sulfate appears. Compare that with where the atomic mass of the iron appears. Can you formulate a general rule about how and when to use atomic mass versus molar mass? Think about this a little before reading on.

Let's focus on multi-step versus single-step conversion problems. Your text has a nice discussion about the meaning of the word *convert* (page 275). Which of the following two statements is true?

1. Converting means a chemical transformation of one element or compound into another.
2. Converting means changing the unit system we are using to refer to the element or compound.

Decide which one is true before reading below. Do not peek!

If you chose (1), then go back and reread the section on converting on page 275. It is important to realize that the conversion process is just an exercise we do on paper. We can measure the length of the room in which you are sitting in meters, feet, yards, inches, and centimeters as well as a variety of other silly units. The point is, *has the length of the room changed as we change the unit system?* Clearly not. Changing unit systems is the process of converting, and it has no impact on the amount of stuff we have. In chemistry, we convert between grams and moles of a single substance all the time. We also need to convert between grams and moles of different substances. These are two related but different types of conversions.

Converting between grams and moles of a single substance is a *single-step* conversion.

Converting between grams and moles (or grams and grams) of two *different* substances is a multistep conversion.

For the following multistep conversions, fill in the missing blanks:

1. Determine the number of grams of the compound $C_6H_3N_7O_3$ that contain 2.1 grams of carbon.

$$2.1 \text{ g of C} \times \frac{1 \text{ mole of C}}{\underline{\hspace{2em}} \text{ g of C}} = 0.17 \text{ moles of C}$$

$$0.17 \text{ moles of C} \times \frac{1 \text{ mole of } C_6H_3N_7O_3}{\underline{\hspace{2em}} \text{ moles of C}} = 0.029 \text{ moles of } C_6H_3N_7O_3$$

$$0.029 \text{ moles of } C_6H_3N_7O_3 \times \frac{\underline{\hspace{2em}} \text{ g of } C_6H_3N_7O_3}{1 \text{ mole of } C_6H_3N_7O_3} = \underline{\hspace{2em}} \text{ g of } C_6H_3N_7O_3$$

How many moles of $C_6H_3N_7O_3$ contain 2.1 grams of carbon? (Do not do any new calculations; just read your answer from the steps above.)

Notice that these problems are set up by starting with the information you have and working your way, step by step, to the desired answer.

2. How many grams of carbon are there in 10.2 grams of $C_6H_3N_7O_3$?

$$10.2 \text{ g of } C_6H_3N_7O_3 \times \frac{1 \text{ mole of } C_6H_3N_7O_3}{\underline{\hspace{2em}} \text{ g of } C_6H_3N_7O_3} = \underline{\hspace{2em}} \text{ moles of } C_6H_3N_7O_3$$

$$\underline{\hspace{2em}} \text{ moles of } C_6H_3N_7O_3 \times \frac{\underline{\hspace{2em}} \text{ moles of C}}{1 \text{ mole of } C_6H_3N_7O_3} = \underline{\hspace{2em}} \text{ grams of C}$$

$$\underline{\hspace{2em}} \text{ moles of C} \times \frac{\underline{\hspace{2em}} \text{ g of C}}{1 \text{ mole of C}} = \underline{\hspace{2em}} \text{ g of C}$$

3. Practice the following until you feel very certain about your ability to do these types of multistep conversions:
   a. Pick a compound. Let's use $C_{\square}H_{\square}N_{\square}O_{\square}$.

      You fill in the boxes with any integers you want. Don't worry about whether your compound is a real chemical; just make up the numbers.
   b. Design a problem by picking numbers. Fill in the blanks—just make up a number for grams, but pick an appropriate atom:

      Determine the number of grams of your compound that contain ___ grams of ___.
   c. Solve the problem.
   d. Design a slightly different problem. Fill in the blanks below:

      How many grams of _______ are there in _______ grams of your compound?
   e. Repeat a–d until you are 100% confident you could work any problem of this type!

Practice 4 Using Density and Volume to Calculate Grams of a Compound

A minor modification of the above is to use density and volume to compute grams. This is particularly useful when the substance is a liquid.

Worked Example: How many moles of benzene (C_6H_6) are there in 23.2 mL of benzene? The density of benzene is 0.88 g/mL.

Thinking through this problem:
I know that once I get grams of benzene, I can easily get moles because I am extremely confident about the gram-to-mole conversion. So all I need is grams of benzene:

$$23.2 \text{ mL benzene} \times \frac{0.876 \text{ g of benzene}}{1 \text{ mL of benzene}} = 20.4 \text{ g of benzene}$$

From the above, calculate the moles of benzene (C_6H_6). In fact, go one step further and calculate the grams of carbon in this volume of benzene. To do this, write down a road map through the problem:

ROAD MAP

volume benzene → grams benzene → moles benzene → moles carbon → grams carbon

Notice that the density calculation adds one more calculation to your multistep algorithm.

Try another benzene calculation. Let's put all the steps together in one problem.

1. How many grams of hydrogen are there in 50.0 mL of benzene?

   Think your way through this five-step problem. Write down the road map first, and then work your way through it.

Try It Yourself:

2. How many grams of oxygen are there in 28.4 mL of ethanol (C_2H_4O)? The density of enthanol is 0.789 g/mL.
3. How many grams of hydrogen are there in 4.00 L of octane (C_8H_{18})?

Practice 5 Mass Composition and Percentages

The gram-to-mole conversions are extremely central to the quantitative manipulations in chemistry. However, chemists also think about the grams and mass percentages of compounds. First, let's be clear about percent composition.

1. Consider methanol (CH_4O). Compute the number of grams of carbon in each of the following cases:
   a. 32.04 grams of methanol
   b. 10.2 grams of methanol
   c. 51.2 grams of methanol

Next, fill in the following table:

Grams of Methanol	Grams of Carbon	Percent Carbon
32.04 grams		
10.2 grams		
51.2 grams		

What have you discovered about the percent carbon?

Percentage is a *relative* measure. That means we always have the same percentage no matter how much stuff we have. Grams of carbon, however, is not a relative measure. Thus, the grams of carbon do vary with the amount of material. What is the difference between an absolute amount, such as grams, and a relative amount, such as percent?

Because percent is a relative amount, the amount of material does not have to be specified when a percent is asked. For example, I can ask what is the percent carbon in methanol without ever specifying how much methanol. So how would you compute the percent? Look at the table above. Can you choose any amount of methanol?

Since we are free to choose any amount of methanol, let's pick a useful amount—the number of grams in one mole. Fill in the following table:

Grams of Methanol in One Mole (*What is this called?*)	Mole of Carbon in One Mole of Methanol	Grams of Carbon in One Mole of Methanol	Percent Carbon in Methanol

We can make a table like this for every percent composition problem. Let's modify the table and ask what is the percent composition of all elements in methanol. We can make the table the following way:

Percent Composition Table

Molar mass of methanol	
Moles of carbon in one mole of methanol	
Moles of hydrogen in one mole of methanol	
Moles of oxygen in one mole of methanol	
Grams of carbon in one mole of methanol	
Grams of hydrogen in one mole of methanol	
Grams of oxygen in one mole of methanol	
Percent carbon (by mass) in methanol	
Percent hydrogen in methanol	
Percent oxygen in methanol	

Work on problem 24 (page 287) but modify it to compute the percent composition of all the elements in each of the compounds.

Practice 6 Using Percent Composition

Once we know the percent composition, we can use it to compute absolute amounts. Reconsider the earlier table:

Grams of Methanol	Grams of Carbon	Percent Carbon
32.04 grams		
10.2 grams		
51.2 grams		

Because we now know that the percent carbon in methanol is 37.49%, use this to compute grams of carbon. When you're done, you can check your answer with your earlier calculations.

Worked Example: From the percent, we know to interpret the meaning of 37.49% as 37.49 grams of carbon per 100 grams of methanol:

$$\frac{37.49 \text{ grams of carbon}}{100 \text{ grams of methanol}} = \frac{x \text{ grams of carbon}}{32.04 \text{ grams of methanol}}$$

You compute x and compare it with your earlier calculations.

Work on problem 27. Notice that nowhere in the problem is the word "percent" used. Rephrase the problem so that the word percent is included in the sentence. Then work the problem setting up a percent composition table.

Practice 7 Empirical Formulas

Critical to the entire study of chemistry is the chemical formula. Ever wonder how chemists figure out the chemical formula of compounds? The method is fairly straightforward. The amount (in grams) of the individual elements is measured. From the grams, moles of the individual elements are obtained. Once we know the moles, we know the mole ratios. Then it is a simple matter of converting all the numbers to integers. Fill in the following table:

A 0.50-gram sample containing carbon, hydrogen, and nitrogen was analyzed. The following masses were measured.

Element	Mass Measured	Moles
C	0.261	
H	0.094	
N		

The empirical formula is obtained from the above data. Let's look at the table with the data filled in:

Empirical Formula Table

Element	Mass Measured (g)	Moles	Moles in Integers
C	0.261	**0.0217**	**0.0217/0.0108 = 2.01 ~ 2**
H	0.087	**0.0863**	**0.0863/0.0108 = 7.99 ~ 8**
N	**0.151**	**0.0108**	**0.0108/0.0108 = 1**

The empirical formula is read from the last column: C_2H_8N

We can use this table to work almost any empirical formula problem. For example, use the table to obtain the empirical formula in Example 8.19 (page 289).

Element	Mass Measured (g)	Moles	Moles in Integers
N		**1.40**	
H		**2.80**	

Since we are given moles, this problem is easier than the one before because we do not need to convert grams to moles.

Occasionally, we are given numbers of atoms instead of mass. We can modify the table to include this possibility (See Example 8.20, page 289):

Element	Mass (g) or Atoms	Moles	Moles in Integers
C	5.06×10^{23} atoms		
H	1.012×10^{24} atoms		

The last possibility is to be given the mass in percentages instead of grams. Don't be fooled by this! Because it is a percentage, you can easily convert to grams by imagining you have 100 grams of the compound. If the compound is 34% carbon, for example, you will have 34 grams.

The table will help you organize the data. Try working empirical formula problems 39, 40, and 41 using the table.

Practice 8 Molecular Formulas (Chemical Formulas)

The molecular formula is obtained from the empirical formula. The hardest part is getting the empirical formula (which really isn't so hard). If you know the molar mass and you know the empirical formula, you can then determine the molecular formula. Notice that every molecular formula problem looks like an empirical formula problem except that the molar mass is given. For example, try problem 54 in your text.

CHAPTER 9

Chemical Stoichiometry

Checklist

What you need to be able to do when you finish Chapter 9

- Use the balanced chemical equation to do mole/mole calculations.
- Use the balanced chemical equation to do mass/mole calculations.
- Use the balanced chemical equation to do mass/mass calculations.
- Identify a limiting reactant.
- Compute a theoretical yield.
- Compute a percent yield.

Practice Problems and Study Hints

Chapter 9 develops the core skill in quantitative chemistry: computing mass changes due to a chemical reaction. All the work you have done so far will be used in solving stoichiometry problems.

Practice 1 Mole/Mole and Mole/Mass Conversions

There is a central algorithm you can keep in mind anytime you do stoichiometry problems. The algorithm is:

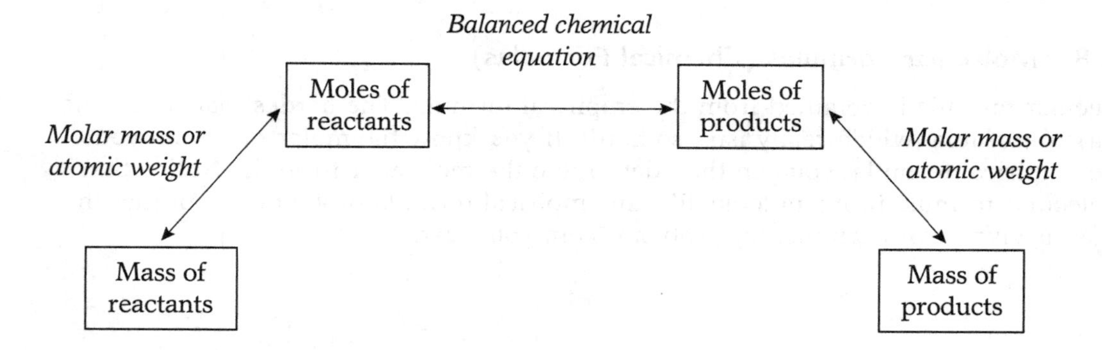

The double arrows mean you can go either way. The information you need to go from one box to the next is in italic font above the arrow.

Your book divides this algorithm into two sections: mole-mole calculations and mole-mass calculations. This is a good way to practice each of the pieces of the algorithm. For this Study Guide, though, I will organize the two pieces into one section.

One way to organize the information is to make a table that will contain *all* the information you might need to solve the problem. Your task is to fill in all the blanks of the table.

A sample table for the reaction:

$$CH_4 + 2O_2 \rightarrow CO_2 + 2H_2O$$

is below. It is given that you start with 5.0 grams of methane (CH_4) and excess oxygen.

Fill in the table below with everything known before doing any calculations:

Reactant (Molar Mass)	Reactant Mass	Reactant Moles	Product Moles	Product Mass	Product
CH_4 (16)	5.0 g				CO_2 (44)
O_2 (32)	Excess				H_2O (18)

Begin to fill in the table—convert all grams to moles.

Reactant (Molar Mass)	Reactant Mass	Reactant Moles	Product Moles	Product Mass	Product
CH_4 (16)	5.0 g	**0.31**			CO_2 (44)
O_2 (32)	Excess				H_2O (18)

When we have an excess of one reactant, the problem is much easier. We do not need to worry about the excess reactant, we know that we have more oxygen than we need. All we need to do is compute how much oxygen will react with the methane.

Because we are converting between two different compounds, we need to use the coefficients in the balanced chemical equation:

$$\frac{2 \text{ moles oxygen}}{1 \text{ mole methane}} = \frac{x \text{ moles oxygen}}{0.31 \text{ mole methane}}$$

From balanced chemical equation

Solving for x, we obtain: $2(0.31) = x$; $x = 0.62$ moles oxygen. Let's add this to the table:

Reactant (Molar Mass)	Reactant Mass	Reactant Moles	Product Moles	Product Mass	Product
CH_4 (16)	5.0 g	**0.31**			CO_2 (44)
O_2 (32)	Excess	**0.62**			H_2O (18)

Notice that whenever we go from one compound to another, we can use *only* the moles. The key to this is the central mole boxes highlighted above. We can fill in these central boxes once we know *one* box.

Let's fill in the remaining two boxes. Remember that whenever we go from one compound to another we have to use the balanced chemical equation.

Methane to carbon dioxide:

$$\frac{\text{1 mole carbon dioxide}}{\text{1 mole methane}} = \frac{x \text{ moles carbon dioxide}}{\text{0.31 moles methane}}$$

x moles carbon dioxide = 0.31

When the mole/mole ratio is 1 to 1, you may not need to set up the proportion. But let's keep setting up the proportions just to see that there is a system to these problems.

Lastly, let's get moles of H_2O. Notice that it doesn't matter which other compound you use, because you will get the same answer. Simply pick any box.

I am going to pick moles of oxygen to moles of water.

$$\frac{\text{2 moles oxygen}}{\text{2 moles water}} = \frac{\text{0.62 moles oxygen}}{x \text{ moles water}}$$

The moles of water = 0.62. Again, here is a case where it might be obvious because the oxygen-to-water mole ratio is 1 to 1.

Filling in the table with our calculated mole values gives:

Reactant (Molar Mass)	Reactant Mass	Reactant Moles	Product Moles	Product Mass	Product
CH_4 (16)	5.0 g	**0.31**	**0.31**		CO_2 (44)
O_2 (32)	Excess	**0.62**	**0.62**		H_2O (18)

It should be straightforward to convert moles of products to mass of products. For now, you can check my calculations:

Reactant (Molar Mass)	Reactant Mass	Reactant Moles	Product Moles	Product Mass	Product
CH_4 (16)	5.0 g	**0.31**	**0.31**	13.6 ~ 14 g	CO_2 (44)
O_2 (32)	Excess	**0.62**	**0.62**	11.1 ~ 11 g	H_2O (18)

The answers are computed to two significant figures.

Look at this table and realize that you can answer all of the following questions related to the reaction of methane with oxygen to form carbon dioxide and water, given that you started with 5.0 grams of methane and excess oxygen:

a. How many grams of carbon dioxide were formed?

b. How many grams of water were formed?

c. How many moles of oxygen reacted?

d. How many moles of carbon dioxide were formed?

e. How many moles of water were formed?

Although it is not in the table, you can use the table to answer the question: "How many grams of oxygen reacted?"

You may wonder why you are calculating quantities for this table that are not related to the question that was asked. This is a reasonable question and, if you use the table, you will sometimes calculate quantities that you didn't need. But just think how much practice with stoichiometry problems the table will give you! In order to be successful in this course, you *must* be able to complete stoichiometry problems. In addition, it will also help you understand what is being asked in the questions.

Try using the table to do problems 7, 10, and 12 on page 312. Even though these problems do not ask you to do any *mass* calculations, calculate the mass anyway. **Fill in all the blocks of the table!**

Skip right to problems 16 to 19 on page 320.

These kinds of calculations are so important that you should work more problems until you feel completely comfortable with the calculations. Do not skip steps until you can fill in the table without making any mistakes.

Unless you feel completely comfortable and ready to move on, work all the problems on page 320 using the table.

Practice 2 Expanding the Stoichiometric Table: Limiting Reactants

You should now be a master at basic stoichiometry. Realize that because we can control how much of our reactants with which we start, we may not always start with the correct *stoichiometric* amount. For example, consider the equation for the combustion of methane:

$$CH_4 + 2O_2 \rightarrow CO_2 + 2H_2O$$

This equation tells us that methane reacts with oxygen in a 1-to-2 mole ratio. You can certainly imagine putting methane and oxygen in a container in any mole ratio you desired. For example, suppose you put 1 mole of methane and 3 moles of oxygen in the container. The 1 mole of methane will react with 2 moles of oxygen because that is the *combining* ratio. When 1 mole of methane reacts with 2 moles of oxygen, we have 1 mole of oxygen left over.

We have to keep track of what reacts and what is left over. But before we do that, we have to figure out which reactant is the *limiting reactant.* In the example above, the methane was the limiting reactant because the limiting reactant is the one that is used up. The oxygen is then called the *excess reactant.*

We can practice limiting reactant type problems using the table method. Look at problem 31 on page 330. The balanced chemical equation given is:

$CaO(s) + 3C(s) \rightarrow CaC_2(s) + CO(g)$

You start with 46.0 grams of calcium oxide (CaO) and 28.4 grams of carbon. Make *two* tables—one as if the CaO were the limiting reactant (the carbon was excess) and one with the carbon as the limiting reactant (calcium oxide is excess).

Calcium oxide is limiting.

Reactant (Molar Mass)	Reactant Mass	Reactant Moles	Product Moles	Product Mass	Product
CaO (56.1)	46.0 g	**0.820**	**0.820**		CaC_2 (64.1)
C (12.011)	Excess	**2.46**	**0.820**		CO (28.0)

Cover the table below. Make the same table for the scenario that C is limiting. Do not peek at the answers in the next table.

Carbon is limiting.

Reactant (Molar Mass)	Reactant Mass	Reactant Moles	Product Moles	Product Mass	Product
CaO (56.1)	Excess	**0.788**	**0.788**		CaC_2 (64.1)
C (12.011)	28.4 g	**2.36**	**0.788**		CO (28.0)

Look at the two tables and decide which one represents what really happens when you mix 28.4 grams of C and 46.0 grams of CaO. The table that shows the *least* amount of product made is the correct table. Therefore, for the reaction above, we would identify carbon as the limiting reactant.

Ponder these two tables for a moment and make sure the logic is clear. With limiting reactant problems, we will always have to do two calculations. Let's do these two calculations by making two tables, but we will modify the procedure above.

Construct the table but put the masses of both reactants in the table. In our example above, the table would look like

Reactant (Molar Mass)	Reactant Mass	Reactant Moles	Product Moles	Product Mass	Product
CaO (56.1)	46.0 g				CaC_2 (64.1)
C (12.011)	28.4				CO (28.0)

Calculate the reactant moles.

Reactant (Molar Mass)	Reactant Mass	Reactant Moles	Product Moles	Product Mass	Product
CaO (56.1)	46.0 g	**0.820**			CaC_2 (64.1)
C (12.011)	28.4	**2.36**			CO (28.0)

Next, identify one of the product moles that you will use as a test calculation. It can be either one. Circle the one you will use.

Reactant (Molar Mass)	Reactant Mass	Reactant Moles	Product Moles	Product Mass	Product
CaO (56.1)	46.0 g	**0.820**			CaC_2 (64.1)
C (12.011)	28.4	**2.36**			CO (28.0)

Compute the moles of this product twice—once using the CaO and once using the C.

Reactant (Molar Mass)	Reactant Mass	Reactant Moles	Product Moles	Product Mass	Product
CaO (56.1)	46.0 g	**0.820**	**0.820** / **0.788**		CaC_2 (64.1)
C (12.011)	28.4	**2.36**			CO (28.0)

Identify the *smaller* one as the limiting reactant. Highlight the limiting reactant in the table and cross out the excess reactant:

Reactant (Molar Mass)	Reactant Mass	Reactant Moles	Product Moles	Product Mass	Product
CaO (56.1)	46.0 g	~~**0.820**~~	~~**0.820**~~ / **0.788**		CaC_2 (64.1)
C (12.011)	28.4	**2.36**			CO (28.0)

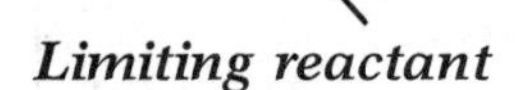

At this point, we need to rewrite the table using just the limiting reactant information above. We also need to add an additional column.

Reactant (Molar Mass)	Reactant Mass	Reactant Mass Left Over	Reactant Moles	Product Moles	Product Mass	Product
CaO (56.1)	46.0 g			**0.788**		CaC_2 (64.1)
C (12.011)	28.4		**2.36**			CO (28.0)

Notice that I left out the results of the excess reactant calculations. You should only include the moles from the limiting reactant calculation. Next, let's fill in all the boxes except the reactant mass left over. You should be able to easily do this now because you already found the limiting reactant; therefore, the problem is simply a basic stoichiometry problem. Fill in the boxes and check my answers.

Reactant (Molar Mass)	Reactant Mass	Reactant Mass Left Over	Reactant Moles	Product Moles	Product Mass	Product
CaO (56.1)	46.0 g		**0.788**	**0.788**	**50.5 g**	CaC_2 (64.1)
C (12.011)	28.4		**2.36**	**0.788**	**22.1 g**	CO (28.0)

What about mass left over from the reactants? Obviously, the limiting reactant gets all used up, therefore, there is no limiting reactant left over. But there is reactant left over that comes from the excess reactant. Let's compute how much.

We calculated **0.820** moles in our limiting reactant table. We reacted **0.788 moles** from the table above. Therefore, we have:

$$0.820 - 0.788 = 0.032 \text{ moles left over.}$$

Convert these moles to mass so we can fill in the rest of the table.

$$0.032 \text{ moles} \times \frac{56.1 \text{ grams}}{\text{mole}} = 1.80 \text{ grams}$$

Reactant (Molar Mass)	Reactant Mass	Reactant Mass Left Over	Reactant Moles	Product Moles	Product Mass	Product
CaO (56.1)	46.0 g	**1.80 g**	**0.788**	**0.788**	**50.5 g**	CaC_2 (64.1)
C (12.011)	28.4	**0.00 g**	**2.36**	**0.788**	**22.1 g**	CO (28.0)

Using this table, look at the problem 31 on page 330. Answer the questions.

I hope you noticed that answering the questions from the table is easy. All the hard work goes into constructing the table. Because I have shown each step, the table construction seems more cumbersome than it really is. Let's combine steps and make just two tables.

Worked Example: Consider this problem.

$$3NO_2(g) + H_2O(l) \rightarrow 2HNO_3(aq) + NO(g)$$

Suppose I start with 5.23 g of NO_2 and 12.5 g of H_2O. Find the limiting reactant and all the masses of all the compounds left in the reaction vessel.

To solve this problem we will have to construct two tables: a limiting reactant table, and a reaction table.

The Limiting Reactant Table

Reactant (Molar Mass)	Reactant Mass	Reactant Moles	Product Moles	Product Mass	Product
NO_2 (46.0)	5.23 g	**0.114**			HNO_3 (63.0)
H_2O (18.0)	12.5 g	**0.694**	0.038 / 0.694		NO (30.0)

The Reaction Table

Reactant (Molar Mass)	Reactant Mass	Reactant Mass Left Over	Reactant Moles	Product Moles	Product Mass	Product
NO_2 (46.0)	5.23 g	**0.00**	***0.114***	**0.076**	**4.79 g**	HNO_3 (63.0)
H_2O (18.0)	12.5 g	**11.8 g**	**0.038**	***0.038***	**1.14 g**	NO (30.0)

(The numbers in ***bold italic*** I brought down from the limiting reactant table. The **bold** numbers are calculated in the reaction table.)

Now given the table, let's ask a series of questions referring to the reaction given that we mixed 5.23 grams of nitrogen dioxide with 12.5 grams of water.

$$3NO_2(g) + H_2O(l) \rightarrow 2HNO_3(aq) + NO(g)$$

a. What was the limiting reactant?

b. How much of the excess reactant was left over?

c. How much of the excess reactant reacted? (You have to calculate this).

d. How many grams of NO can you make?

e. How many grams of HNO_3?

Now you practice limiting reactant problems by making the tables. Do problems 33–36 in your book (page 332). Make both a limiting reactant and a reaction table and fill in all the blanks. Then, answer the questions in the text.

Practice 3 Percent and Theoretical Yield

It is very easy to do percent and theoretical yield once we realize that the reaction table is for a *100%* yield. Thus, when we need to calculate theoretical yield, we still have to make a reaction table (and a limiting reactant table).

We also have to realize that as long as we are given a mass in the reaction table, we can compute all the other quantities. This means we can also work backward. Working backward, in fact, is easier because this is not a limiting reactant problem.

For example, suppose I want to make 12.5 grams of ammonia from nitrogen and hydrogen. The balanced chemical equation is:

$N_2 + 3H_2 \rightarrow 2NH_3$

I realize that this is not a limiting reactant problem, therefore, I only need a reaction table.

Reaction table for $N_2 + 3H_2 \rightarrow 2NH_3$.

Reactant (Molar Mass)	Reactant Mass	Reactant Mass Left Over	Reactant Moles	Product Moles	Product Mass	Product
N_2 (28.0)		NA			12.5 g	NH_3 (17.0)
H_2 (2.02)		NA				

Fill in all the relevant blank cells. Answer the question, "How much of each reactant will you need to make 12.5 grams of ammonia?"

Once we realize that we can move around the table forward or backward we are ready to add percent yield into the calculation. Remember, in order to solve any stoichiometry problem, we need only know masses: either masses of reactants or masses of products.

A. Working a Problem Forward

Suppose I react 10.0 grams of nitrogen with 10.0 grams of hydrogen. I find that I make 2.0 grams of ammonia. What is the percent yield of the reaction?

To answer this we first must realize that this is a limiting reactant problem. (**Hint:** It is always a limiting reactant problem when we are given reactants!)

Limiting Reactant Table for $N_2 + 3H_2 \rightarrow 2NH_3$.

Reactant (Molar Mass)	Reactant Mass	Reactant Mass Left Over	Reactant Moles	Product Moles	Product Mass	Product
N_2 (28.0)	10.0	NA	0.357	0.714 / 3.30		NH_3 (17.0)
H_2 (2.02)	10.0	NA	4.95			

Reaction table for $N_2 + 3H_2 \rightarrow 2NH_3$.

Reactant (Molar Mass)	Reactant Mass	Reactant Mass Left Over	Reactant Moles	Product Moles	Product Mass	Product
N_2 (28.0)	10.0		**0.357**	**0.714**	**12.1 g**	NH_3 (17.0)
H_2 (2.02)	10.0		**1.07**			

Let's keep track of percent yield by adding three lines to the table:

Percent yield: __________

Theoretical yield: __________

Actual yield: __________

We can fill in the theoretical yield from the reaction table: 12.1 g. The actual yield is 2.0 grams of ammonia. Thus, I need to calculate the percent yield:

Percent yield: UNKNOWN

Theoretical yield: 12.1 g

Actual yield: 2.0 g

$$\text{percent yield} = \frac{\text{actual yield}}{\text{theoretical yield}} \times 100 = \frac{2.0\text{ g}}{12.1\text{ g}} \times 100 = 17\%\text{ yield}$$

Stoichiometry is really one big accounting problem!! Making tables and filling in the boxes will help you stay organized. It also helps you develop a feel for these types of problems. With experience, you will become faster at doing these problems and will be able to skip steps.

B. Working the Problem Backward

Suppose in the previous problem, I was asked to find the amount of nitrogen and hydrogen I would need to make 6.23 grams ammonia given that the percent yield of the reaction is 17%.

To solve this problem, I realize that I am working backward and can therefore go directly to my reaction table. This time, I am going to include the yield lines.

Reaction table for $N_2 + 3H_2 \rightarrow 2NH_3$.

Reactant (Molar Mass)	Reactant Mass	Reactant Mass Left Over	Reactant Moles	Product Moles	Product Mass	Product
N_2 (28.0)		**NA**				NH_3 (17.0)
H_2 (2.02)		**NA**				

Percent yield: 17%

Theoretical yield: ______

Actual yield: 6.23 grams

Because the table is entirely blank, I can't do anything until I can get a mass into the table. From the yield calculations I can determine the theoretical yield. It is the theoretical yield that *always* appears in the table.

$$\text{percent yield} = \frac{\text{actual yield}}{\text{theoretical yield}} \times 100$$

$$17\% = \frac{6.23 \text{ g}}{x} \times 100$$

solving for x we get: $x = 6.23/0.17 = 36.6$ g

Given that my theoretical yield is 36.6 grams, I will put this in the table and solve for all the unknown quantities. You can easily do this calculation. (I get 30.1 grams of nitrogen.)

Reaction table for $N_2 + 3H_2 \rightarrow 2NH_3$.

Reactant (Molar Mass)	Reactant Mass	Reactant Mass Left Over	Reactant Moles	Product Moles	Product Mass	Product
N_2 (28.0)		NA			**36.6 g**	NH_3 (17.0)
H_2 (2.02)		NA				

Percent yield: 17%

Theoretical yield: 36.6 grams

Actual yield: 6.23 grams

Using the tables practice working the problems on pages 333 and 334.

Try It Yourself: Add the following calculations to these problems:

#50 *Add:* Suppose it is known that the reaction has a yield of 30.2%. How much NO will you make in this case?

#51 *Add.* If 1.35 g of NO_2, what is the percent yield of the reaction?

#52 *Change it to read:* If I want to produce 2.5 kg (kilograms!) of $Na_2C_2O_4$ and I know the percent yield of the reaction is 45%, then how many kilograms of $C_6H_{10}O_4$ must I start with?

CHAPTER 10

Discovering the Gas Laws

Checklist

What you need to be able to do when you finish Chapter 10

- Graph data, determine scale, label axes and title.
- Obtain slope of straight line plots, use the slope-intercept equation to write the equation of the straight line.
- Be able to convert among pressure units: atm, mmHg, torr.
- Graph PV data, graph the data to generate the appropriate linear relationship.
- Calculate pressure and volume changes at constant temperature.
- Graph volume vs. temperature (in Celsius) for a gas and know the relationship of this graph to the Kelvin scale.
- Convert between Celsius and Kelvin.
- Graph V and T data and graph the data to generate the appropriate linear relationship.
- Compute changes in volume and temperature at constant pressure.
- Put all the variables together into the ideal gas law—know the units of the ideal gas law.
- Use the ideal gas law to compute changes in pressure, volume, temperature, moles.
- Use the ideal gas law in gas stoichiometry.
- Articulate the model of the gas that is consistent with the ideal gas law.

Practice Problems and Study Hints

Practice 1 Graphing Data

This is an excellent opportunity to practice your graphing skills. You should be able to do many things with graphs, including: graphing data, generating the slope-intercept equation for the data, and predicting values. Before doing these exercises, practice graphing with arbitrary (i.e., not gas law) data. Be sure you can do the problems on pages 345 and 347. In particular, do problems 2 to 4 and then do problem 5. In problem 5, think about how you would get the intercept of the line when you know two points on the line? (Think about what the *x*-coordinate is for the *y*-intercept.)

Last, do problem 9, and then you are ready to try to the following.

1. *Graphing PV data.*

Below are PV data for 0.1 mole of argon at room temperature (25 °C). Graph the data with P as the dependent variable and 1/V as the independent variable. Before graphing, always: 1) generate a table with the data you will graph; 2) decide on an appropriate scale to fill the graph, and 3) draw and label your axes.

P (torr)	V (L)	1/V (L^{-1})
930	2	
372	5	
232	8	

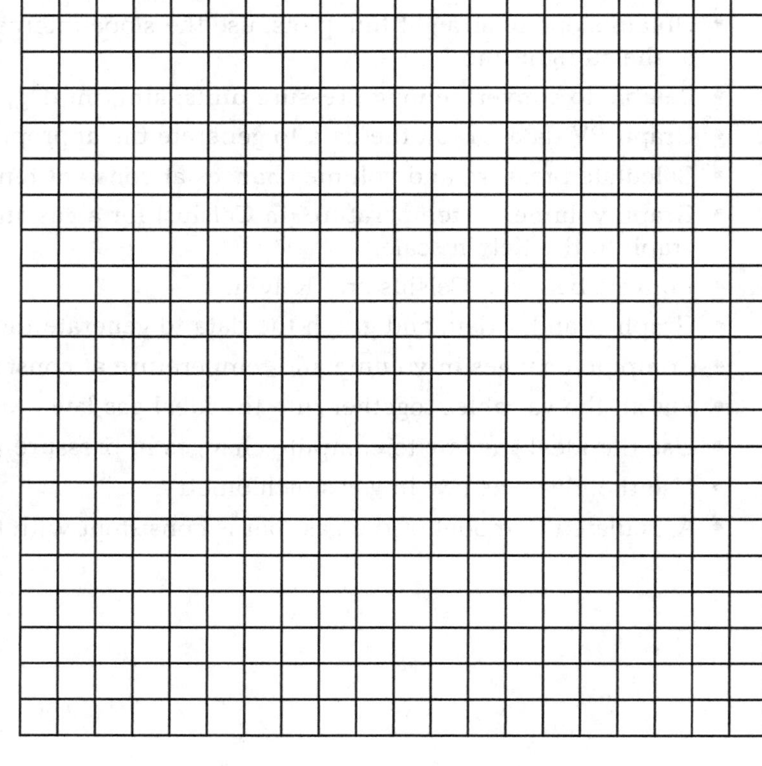

Using your graph, determine the following:

- The slope of the straight line
- the y-intercept
- the slope-intercept equation for the data
- Predict the value of the pressure with the volume is 4 L. (Be sure to compute 1/V first.)

2. *Reading PV graphs*
   - Look at your graph again. Be sure you understand how the graph looks. Answer the following questions based on the graph or the equation you generated.
   - Is the slope of the graph positive or negative?
   - As 1/V gets bigger, what happens to V? Does it get bigger or smaller?
   - As 1/V gets bigger, what happens to the pressure?
   - As V gets bigger, what happens to the pressure?
   - How do you read increasing V on your graph? Does it go from left to right on the *x*-axis or from right to left?
   - What best describes the relationship between P and V: linear, inverse, or directly proportional?

3. *Graphing VT data*
   Consider the volume (V) and temperature (T) data obtained for 0.2 moles of helium gas at 450 torr. Graph the data assuming that T is the independent variable and V is the dependent variable. Graph the data making sure that the T value is in Kelvin. (Convert your Celsius temperatures to Kelvin.)

V (L)	T (°C)	T (K)
8.3	25	
16.6	323	
24.8	621	

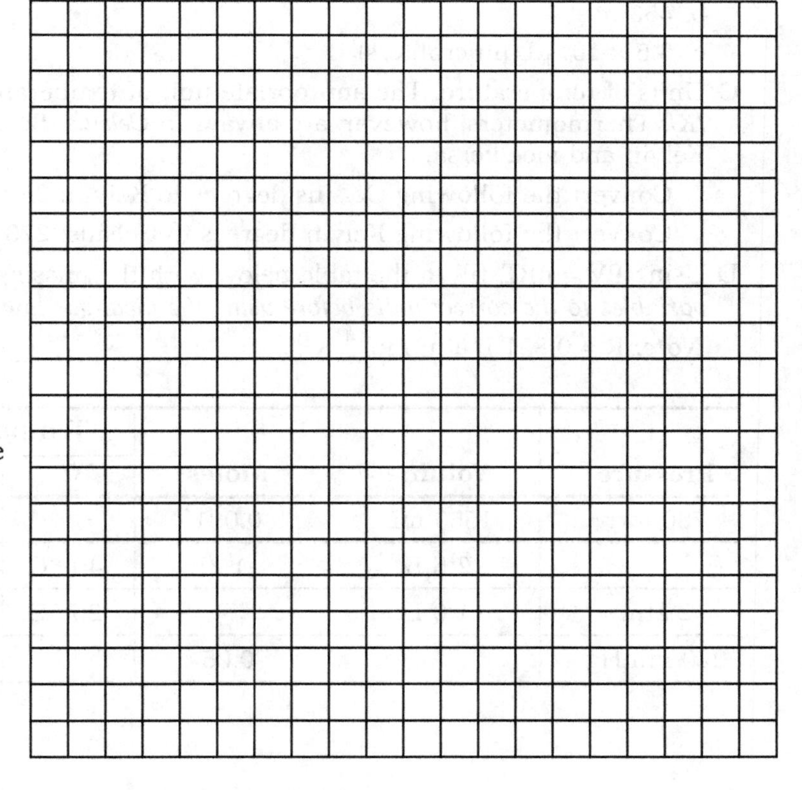

Using your graph, determine the following:
   - The slope of the straight line
   - The *y*-intercept
   - The slope-intercept equation
   - Predict the value of the volume when T = 405 °C (Be sure to convert temperature into Kelvin first.)
   - Using your slope-intercept equation, predict the value of the volume when T = 405 °C (*Again, do not forget to convert the temperature to Kelvin!*)

Note: Be sure all of your graphs are carefully labeled—look at the axis and the title. Can you read the graphs? We will come back to them later. In fact, one measure of how careful you are about labeling your graph is your ability to read the graph many days after you have made it. When you look at your graph a week from now, will you know what you plotted? If your answer to this question is not a big "YES," you need to be more careful about how you label your graphs!

Practice 2 Computing Changes in Gas Variable: The Ideal Gas Law

Before computing changes, be sure you are familiar with all the variables that define the state of a gas. There are four variables: P, V, T, and n. The relationship among all four is expressed in the ideal gas law: $PV = nRT$.

Note: The ideal gas law works only when you have the correct units on P, V, and T.

A. Units of pressure. The appropriate unit of pressure in the ideal gas law is *atmospheres (atm)*. Convert the following to atm.

   a. 710 torr

   b. 380 mmHg

   c. 1450 torr

   d. **Now try this:** Convert 29.2 in Hg (**Hint:** Use 2.54 cm = 1 in) to mmHg and then to atm.

B. Units of volume. The appropriate unit of volume in the ideal gas law is *liters (L)*. Convert the following to liters.

   a. 1200 mL

   b. 453 mL

   c. 3.5×10^8 μL (microliters)

C. Units of temperature. The appropriate unit of temperature in the ideal gas law is *Kelvin (K)*. Thermometers, however, are always in *Celsius*. Be able to convert from Celsius to Kelvin and vice versa.

   Convert the following Celsius degrees to Kelvin: 25 °C, 200 °C, −100 °C.

   Convert the following Kelvin degrees to Celsius: 273 K, 0 K, 298 K, 543 K

D. Using $PV = nRT$, fill in the table below with the missing variable. *Be sure to convert the variables to the correct units before using the ideal gas law.*

   Note: $R = 0.821$ L atm mol^{-1} K^{-1}

Pressure	Volume	Moles	Temperature	
			°C	K
780 torr	1550 mL	0.001		
	2.5 L	1.2	40 °C	
2 atm	1.0 L		25 °C	
220 mmHg		0.05		410 K

Practice 3 Using the Ideal Gas Law to Answer Gas Questions

All problems requiring you to solve the ideal gas law can be answered by making a table such as the one above. The only way you can use the ideal gas law is if three of the four variables are given. Notice that you might have to do some conversion to get the numbers in the table. For example, you may have to convert *grams* to *moles*. Try the following problems, making an ideal gas table:

a. A 25.0-g sample of helium is confined to a volume of 4.5 liters at room temperature (25 °C). What is the pressure of the gas?

b. A 2.5-liter volume of gas at 0 °C is found to have a pressure of 1450 torr. How many moles of gas are there in the container?

c. 0.163 grams of a gaseous element is confined to a volume of 1.0 liter at room temperature (25 °C). What is the unknown gas? (**Hint:** Calculate the number of moles of gas and then find the molar mass of the gas.)

Practice 4 Using the Ideal Gas Law to Compute Changes in the Gas Variables

On page 372, there is a "super table" of all the variables needed to solve gas problems. When we want to consider changes in the state of a gas we have to consider the two cases. To keep track of what is changing and what needs to be calculated let's construct a super table and fill it in with all the variables that we know. Designate "1" as one case (it doesn't matter which one) and "2" as the other case. Fill in the super table below for each of the problems that follow. Note that you can (and should) apply the ideal gas law for each case separately.

Problem	P_1	P_2	V_1	V_2	T_1	T_2	n_1	n_2
a.								
b.								
c.								
d.								

a. 0.5 moles of carbon dioxide gas are confined to a fixed volume of 2.2 liters. The temperature of the gas is increased from 0 °C to 20 °C. What happens to the pressure of the gas?

b. 7.5 grams of carbon monoxide are put into a variable volume container. The pressure of the carbon monoxide is 320 torr at 21°C. How does the volume of the container change when the gas is heated to 42 °C at constant pressure?

c. 0.01 moles of neon are put in a neon sign of total volume 2.0 liters. If the pressure of the gas increases from 93 torr to 183 torr, what happens to the temperature of the gas?

d. A fixed 2.0-liter container has oxygen gas pumped into it. At 25 °C, 36 grams of oxygen are pumped into the container. How does the pressure change when another 36 grams of oxygen are pumped into the container?

Practice 5 Using Proportions to Calculate the Gas Variables

Sometimes, we are not given enough information to compute all of the variables in the ideal gas law. For example, suppose we have a fixed amount of an ideal gas, which we are keeping at a fixed temperature. We decrease the volume of the gas from 3 liters to 1 liter. How does the pressure of the gas change if the pressure was initially 0.5 atm?

To solve this problem, let's use our super table:

P_1	P_2	V_1	V_2	T_1	T_2	n_1	n_2

Fill in the table with everything that is known:

P_1	P_2	V_1	V_2	T_1	T_2	n_1	n_2
0.5 atm		3 liter	1 liter				

Because we don't know the temperature or the moles, we cannot fill in the rest of the table. But we need to realize that neither the temperature nor the moles change. Let's put "no change" into the table for temperature and moles:

P_1	P_2	V_1	V_2	T_1	T_2	n_1	n_2
0.5 atm		3 liter	1 liter	No change	No change	No change	No change

To solve for the P_2 (the empty cell), we have to use the appropriate proportionality, or what is sometimes called, the combined gas law (see page 363):

$$\frac{P_1V_1}{n_1RT_1} = \frac{P_2V_2}{n_2RT_2}$$

We can simplify the combined gas law for a given problem by eliminating the variables that do not change. For the above problem, neither the moles nor the temperature change. The proportion then becomes:

$$P_1V_1 = P_2V$$

We can use this to obtain $P_2 = 1.5$ atm.

For each of the problems below, fill in the table and write the correct proportionality that you can use to solve the problem.

1. Argon gas is sealed into a large container. The pressure of the gas is 4 atm at 30 °C. The gas is heated until the pressure doubles. What is the temperature of the gas?

P_1	P_2	V_1	V_2	T_1	T_2	n_1	n_2

The appropriate proportion is: ________

2. Nitrogen (N_2) gas is pumped into a variable volume container until the pressure is exactly 1 atm. The volume of this gas was measured to 10.0 liters. The volume is then decreased to 2.0 liters. What is the pressure of the gas? Assume the temperature remains constant.

P_1	P_2	V_1	V_2	T_1	T_2	n_1	n_2

The appropriate proportion is: ________

3. Helium is confined in a variable volume container at constant pressure. If helium is heated from 10 °C to 50 °C, then how does the volume of the gas change? The initial volume was measured to be 0.5 liters.

P_1	P_2	V_1	V_2	T_1	T_2	n_1	n_2

The appropriate proportion is: ________

4. A constant pressure, constant temperature apparatus is set up. 2.4 grams of helium are pumped into the apparatus until the volume measures 1.0 liter. More helium is pumped in until the volume measures 2.5 liters. How many additional grams of helium were pumped in?

P_1	P_2	V_1	V_2	T_1	T_2	n_1	n_2

The appropriate proportion is: ________

Practice 6 Using Proportions to Compute Changes in Gas Variables

Sometimes, the problems above do not have specific numbers. For example, we could ask how the pressure of a fixed amount of gas would change if we doubled the temperature (in Kelvin) at constant volume. Notice here we state only that the temperature is doubled—we do not know what the initial temperature was. To solve these problems, let's set up the table exactly the same way:

P_1	P_2	V_1	V_2	T_1	T_2	n_1	n_2
		No change	No change			No change	No change

The appropriate proportion is: ________

In this case, the appropriate proportion is:

$$\frac{P_1}{T_1} = \frac{P_2}{T_2}$$

The problem is that we do not know *values* of the initial pressure or temperature. But let's put in the table what we do know:

P_1	P_2	V_1	V_2	T_1	T_2	n_1	n_2
		No change	No change		$2T_1$	No change	No change

We know the temperature doubled—so T_2 has to be twice T_1. Let's put this into the proportion:

$$\frac{P_1}{T_1} = \frac{P_2}{2T_1}$$

We can now cancel T_1 from the two sides of the equation and we have:

$$P_1 = P_2/2$$

Solving for P_2, we have:

$$P_2 = 2P_1$$

We would report that the pressure also doubled.

Try solving these problems using the table and the appropriate proportion.

1. How does the temperature of the gas change when the pressure of the gas is halved (reduced by a factor of 2)? Assume the volume and number of moles remains constant.

P_1	P_2	V_1	V_2	T_1	T_2	n_1	n_2

The appropriate proportion is: ________

2. A sample of helium is put into a container. The volume of the gas is then tripled. What happens to the pressure of the gas? Assume the temperature of the gas remained constant.

P_1	P_2	V_1	V_2	T_1	T_2	n_1	n_2

The appropriate proportion is: ________

3. Argon is pumped into a fixed container until the pressure of the container reaches the desired value. More argon is then pumped in until the pressure doubles. How much argon is pumped in (related to the initial amount)? Assume the temperature remains constant.

P_1	P_2	V_1	V_2	T_1	T_2	n_1	n_2

The appropriate proportion is: ________

The calculations do not get any harder than what is presented here. You should be able to use the table and the appropriate proportion to work any problem. Practice using the table and the appropriate proportion by working the problems in your book (pages 375–377).

Practice 7 Gas Stoichiometry

All stoichiometry requires the computation of moles. Look at the strategy map on page 363. If you have a solid, you usually know how many *grams* of the solid you have. From the grams, you compute the moles. (Review Chapter 9 for how to do this!) For gases, we get the moles from the ideal gas law. This is an extremely important application of the ideal gas law.

All gas stoichiometry problems require you to figure out the number of moles of the gas. Sometimes you have to use the ideal gas law to get the moles. Sometimes you get the moles from the stoichiometry.

Worked Example: Consider the decomposition of CaO:

$2CaO(s) \rightarrow 2Ca(s) + O_2(g)$

Suppose we ask what *volume* of oxygen is collected when we completely decompose 14.0 grams of calcium oxide at 80°C and 1 atm pressure.

Even though we are being asked for volume, *always think moles!* So the question really is, how many moles of oxygen can I get? I can only get the moles of oxygen from the moles of calcium oxide.

Moles of calcium oxide = 14.0 grams/ 56 g mol^{-1} = 0.25 moles

$$\text{Moles of oxygen} = 0.25 \text{ moles calcium oxide} \times \left(\frac{1 \text{ mole oxygen}}{2 \text{ moles calcium oxide}} \right) = 0.125 \text{ moles}$$

Now that I know moles, I can compute the requested volume:

$V = 0.125 \times 0.0821 \times 353/1 = 3.6$ liters

Worked Example: Consider the same decomposition:

$2CaO(s) \rightarrow 2Ca(s) + O_2(g)$

Suppose I produce 4.3 liters of oxygen gas when I decompose an unknown amount of calcium oxide at 100 °C and 760 torr. How much calcium oxide must I have started with?

To solve this problem, *always think moles.* How many moles of oxygen do I have?

Here we know the pressure, volume, and temperature of the gas. Therefore, I can compute the moles:

Moles = 1 atm × 4.3 liters/(0.081 × 373)

Moles of oxygen = 0.14

I can now work backward using the stoichiometry to determine how many moles of calcium oxide must have decomposed and then how many grams of calcium oxide decomposed.

These are the two kinds of gas stoichiometry calculations you have to do. In both cases we can begin to solve the problem by thinking of the moles of gas. Use the following diagram to help you think about these problems:

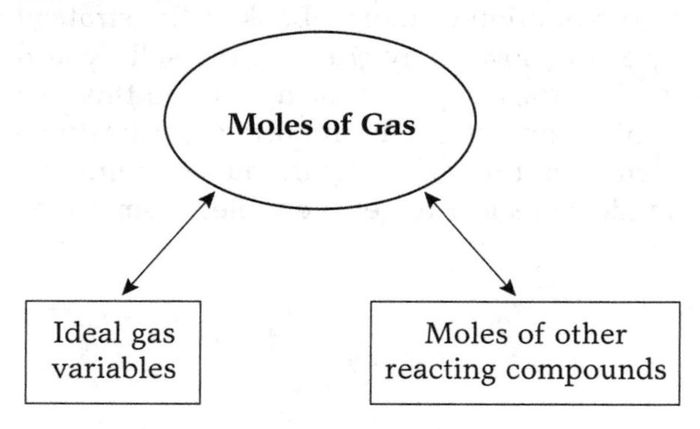

Practice gas stoichiometry by identifying and working all gas stoichiometry problems on pages 375 through 377.

CHAPTER 11

Chemical Systems and Heat

Checklist

What you need to be able to do when you finish Chapter 11

- Be able to make an informed decision about the direction of heat flow.
- Predict the sign (positive or negative) of the heat for the system and surroundings
- Identify heat flow as endo- or exothermic.
- Have a picture in your mind of the relationship between heat and kinetic energy (speed) of a gas.
- Know what a phase change is, and know whether the change requires heat or gives off heat.
- Be able to compute temperature change of a substance when heat is added to or removed from the substance.
- Be able to compute the amount of heat put in or removed from a substance if the temperature change of the substance is measured.

Practice Problems and Study Hints

Practice 1 System, Surroundings, and the Direction of Heat Flow

When discussing heat flow, it is extremely helpful to carefully define the system and surroundings. However, the definition of the system and the surroundings is somewhat arbitrary and applies to the specific situation that you are studying. For example, you might be interested in exploring the flow of heat in a glass of iced tea. In this case, the ice could be the system, and the tea, the glass, and the air in the room could be the surroundings. Or the ice and tea could be the system and the glass and air in the room could be the surroundings. Therefore, it is helpful to be able to get an intuitive feel for the flow of heat in the system and surroundings.

Let's contemplate the iced tea example. First, imagine that you just put the ice in the tea. You probably put the ice in the tea because you wanted to "cool down" the tea. Thus, you imagine heat flowing from the hotter tea to the cooler ice. Let's make a table to keep track of the heat flow.

	Direction of Heat Flow	Endo- or Exothermic?
System: Ice	$q > 0$ (heat flows into the ice)	Endothermic
Surroundings: Tea	$q < 0$ (heat flows from the tea)	Exothermic

Suppose we wait a few minutes and the tea is now cold. We put a lot of ice cubes in the tea so there are still many ice cubes left. We can now think of the ice and tea as the system (notice that they are now at one temperature). The room then becomes the surroundings. Fill in the table for this case.

	Direction of Heat Flow	Endo- or Exothermic?
System: Ice + Tea		
Surroundings: Glass + Room		

Study Example 11.2 in your textbook very carefully. Identify the system and the surroundings for each of the two cases. Then, predict the sign of the heat flow and tell whether the heat flow is endo- or exothermic.

Practice 2 Heat Flow and Phase Changes

In building your intuition, it is very helpful to think about the direction of heat flow in phase changes.

Solid → Liquid Phase Changes

Think about two cases of solid to liquid phase changes. Probably, the most common change is ice to water. (How about melting butter or burning wax?) Consider the system to be the solid and answer the following questions.

- If we want to "melt" the solid, what must we do?
- Is q of the system positive or negative?
- What might be the surroundings?
- Is q of the surroundings positive or negative?
- From the point of view of the system, is the heat flow endo- or exothermic?
- From the point of view of the surroundings, is the heat flow endo- or exothermic?

Liquid → Solid Phase Changes

Next let's consider the reverse. This time freeze the water. Consider the water to be the system. Answer the following questions:

- If we want to "freeze" the liquid, what must we do?
- Is q of the system positive or negative?
- What might be the surroundings?
- Is q of the surroundings positive or negative?
- From the point of view of the system, is the heat flow endo- or exothermic?
- From the point of view of the surroundings, is the heat flow endo- or exothermic?

Diagramming the heat flow

It might be helpful to draw a diagram of the heat flow rather than make a table. We could diagram the solid → liquid phase change as follows:

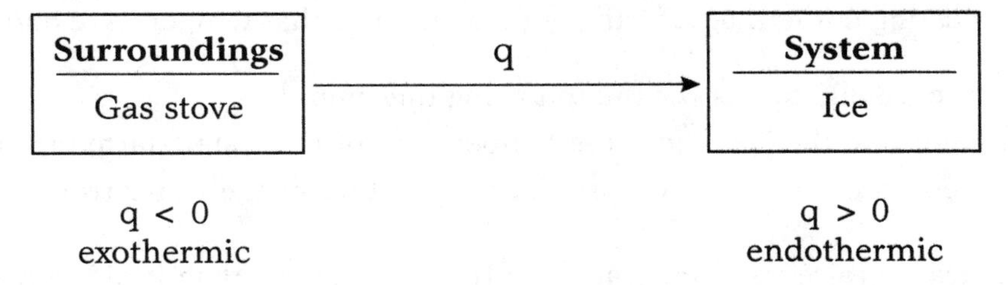

Verbal Description:

Heat leaves the gas stove and flows to the ice (the system). The gas stove loses heat (an exothermic process). The ice gains heat and melts, an endothermic process.

Diagramming the heat flow for freezing liquid water

Diagram the heat flow for freezing liquid water. In your diagram, include the sign of q and the direction of the arrow. Be sure to also give a verbal description of the heat flow.

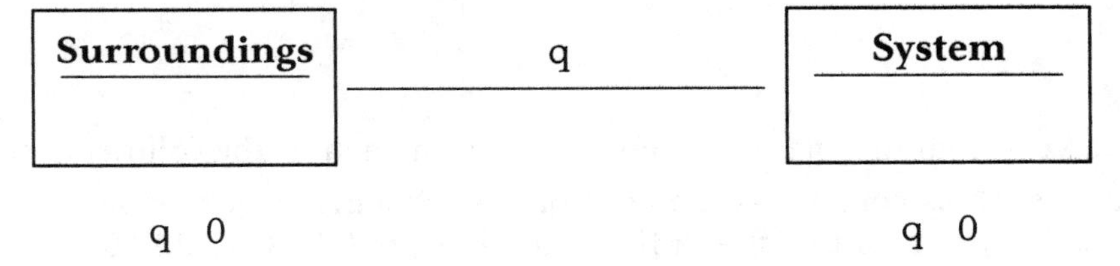

Verbal Description:

Diagramming the heat flow for each process in Table 11-3

Consider the remaining four processes in Table 11-3 of your text. Think of an example for each process and diagram the heat flow. Identify the system, the surroundings, the direction of the heat flow, the sign of the heat for the system and for the surroundings, and give a verbal description of the process.

Practice 3 Temperature and Heat

Our everyday experiences tend to make us think of temperature and heat as synonymous. We talk about "cooling things down" and "heating objects up." Notice how this language makes us think that by putting heat into something it heats up (i.e., its temperature goes up). But there are some examples where putting heat into something does *not* cause its temperature to go up.

Imagine what happens when you put heat into an ice cube. The ice cube melts. While the cube is melting, the temperature of the ice cube and water remains the same. Only when all of the cube has melted will the temperature of the water start to rise. We can graph this to make it clear that the heat and the temperature are not the same. Look at the graph on page 396. This is a heating curve that effectively demonstrates that temperature and heat are *not* synonymous.

Use the graph on page 396 to answer the following questions:

- How much heat went into the ice before the temperature of the water started to rise?
- How much heat went into the water to raise the temperature of the water from 0 °C to 20 °C?
- Did it take more heat to raise the temperature of the water 20 °C than it did to melt the ice?
- Does it make sense to talk about "heating up the ice"? Can you think of a more appropriate descriptive phrase?

Practice 4 Heat and Temperature for a Single Phase

In the case of a single phase, our common sense intuition about the relationship between temperature and heat is put to use. For a single substance (such as water), and a single phase (again, such as water), the heat input or output is related to the temperature rise through the equation:

$$q = cm\Delta T$$

You should become very comfortable using this equation in each of the following settings:

- Given a substance (such as copper), an amount of the substance (such as 5.0 grams), and a temperature rise (5 °C) or an initial and final temperature (20 °C to 25 °C), compute the amount of heat that the copper absorbed.
- Given a substance (such as copper), an amount of the substances, and the amount of heat absorbed by the substance, compute the temperature rise.
- Given a substance, the amount of heat absorbed by the substances, and the temperature rise, compute the mass of the substance.
- Given the heat absorbed by a substance, the mass of the substances, and the temperature rise, compute the specific heat of the substance.

You can try to solve every heat capacity-type problem by making a table with all the variables in it. In these problems you can record temperature in Kelvin or in degrees Celsius. Fill in all the blank cells of the table below. Use Table 11-4, only if needed.

q	c (or substance)	m	$T_{initial}$	T_{final}	ΔT
	Aluminum	50g	25 °C	80 °C	
100 Joules	Copper		33 °C	36 °C	
1200 Joules	Ethanol	100g	25 °C		
800 Joules		100g	18 °C	35 °C	
	Mercury	50g	80 °C	50 °C	-30 °C
-1500 Joules	Mercury	50g	80 °C		
-750 Joules	Gold	10g		30 °C	
	Argon	500g			15 °C
600 Joules		250g	380 K	323 K	

Practice 5 Calculating Heat for a Phase Change

Table 11-5 contains some selected heats of fusion and heats of vaporization. We can use these data to compute the amount of heat needed to melt or evaporate a substance. However, before doing the calculation, we need to be clear on the terminology. Of the four definitions below, which is for the heat of fusion and which is for the heat of vaporization?

a. The heat needed to melt a solid

b. The heat evolved when a liquid freezes

c. The heat needed to evaporate a liquid

d. The heat evolved when a gas liquefies

I hope by now that you recognize that the heat of fusion is the heat needed to melt a solid. Moreover, knowing the heat of fusion means we also know the amount of heat evolved when we freeze a liquid because the heat of freezing is simply the opposite (negative value) of the heat of fusion.

Practice these ideas.

- How much heat is needed to melt 3 moles of ice at 0 °C?
- What is q for the process of freezing 3 moles of water at 0 °C?
- How much heat is needed to boil 50 g of water at 100 °C? (Be sure to convert the grams to moles!)
- What is q for the process of liquefying 50 g of water vapor at 100 °C?
- What is q for the process of evaporating 20 grams of acetic acid at 118.2 °C?
- What is q for the process of liquefying 20 grams of acetic acid at 118.2 °C?

Practice 6 Putting It All Together

We want to be able to compute the amount of heat needed to both change the temperature of a substance and change the phase of the substance. We should now be able to 1) calculate the heat changes (or the final temperatures) and 2) draw heating curves.

Calculating heats and temperatures

Worked Example: How much heat would it take to heat 100 grams of ice at –10 °C to liquid water at 60 °C?

In doing all these problems it is always helpful to think of the process in steps and then to make a table. Remember that each step ends with the change in phase. Here is a sample table:

Phase	T_0	T_f	Mass	Moles	Appropriate Equation or Data	Total Heat
Ice	-10 °C	0 °C	100 g	5.56	$q = cm\Delta T$ (single phase)	$q = 2.03$ $JK^{-1}g^{-1}*100g*10Kdeg$ $q = 2030$ Joules
Ice to Water	0 °C	0 °C	100 g	5.56	$q = q_{fus}*moles$	$q = 6.01\ kJmol^{-1}*5.56$ $q = 33.4$ kJ
Water	0 °C	60 °C	100 g	5.56	$q = cm\Delta T$	$q = 4.18$ $JK^{-1}g^{-1}*100g*60Kdeg$ $q = 25080$ Joules or 25.08 kJ

The total heat needed is the sum of the individual heats: 2.030 kJ + 33.4 kJ + 25.08 kJ = 60.01 kJ.

Notice the following:

- The specific heat capacity is different for each phase. The specific heat capacity of ice is 2.03 $JK^{-1}g^{-1}$. You cannot work this problem unless you know the specific heat capacity for each phase.
- The equation for the heat in the single phase ($cm\Delta T$) requires the mass in *grams.*
- The equation for heat in phase transitions requires the mass of the material in *moles.*
- The single-phase calculation results in an amount of heat in *Joules* whereas the phase transition results in an amount of heat in *kilojoules.*

Now practice computing the total heats for each of the following processes. Make a table for each case.

1. 100 grams of ice is heated from -40 °C to 50 °C.
2. 30 grams of carbon tetrafluoride (liquid) is heated from –150 °C to 50 °C. The boiling point of carbon tetrafluoride is –128 °C. The specific heat of liquid carbon tetrafluoride is 1.16 $Jg^{-1}K^{-1}$. The specific heat of the vapor is 0.68 $Jg^{-1}K^{-1}$ and the heat of vaporization is 11.4 $kJmol^{-1}$.
3. 30 grams of carbon tetrafluoride (vapor) is cooled from 20 °C to –160 °C. How much heat is removed from the carbon tetrafluoride in this process?

Practice 7 Reading a Heating Curve

A graphical representation of the above data clearly demonstrates the relationship between heat and temperature. We can use the table that you have generated to construct a Temperature (°C) vs. Heat (J) graph. Look at the graph on page 397 (Figure 11-8), and answer the following questions:

a. Does adding heat *always* result in a temperature rise?
b. How much heat had to be added to melt all the ice?
c. How much heat had to be added to warm the ice to 0 °C?
d. How much heat would have to be removed to cool the water from 30 °C to 20 °C?

CHAPTER 12

The Atomic Nucleus: Isotopes and Radioactivity

Checklist

What you need to be able to do when you finish Chapter 12

- Define isotope in terms of numbers of protons and neutrons.
- Use the isotope symbol to compute the number of neutrons and protons in the isotope.
- Compute the atomic mass of naturally occurring elements from the atomic masses of each of the isotopes.
- Compute the natural abundances of the individual isotopes from the appropriate data.
- List the three most common types of radiation.
- List the charge and mass number for each of the three most common types of radiation.
- Write balanced nuclear equations using the isotope notation.
- Define half-life, and be able to compute the amount of radioactive material that remains after one, two, three, or more half-lives.

Practice Problems and Study Hints

Practice 1 Reading Isotope Notation

Isotope notation was introduced in Chapter 1. Review the notation and be sure you can compute the number of protons and neutrons from the notation. To practice this, go back and do problems 17 and 18 in Chapter 1 again.

Practice 2 Using Averages to Compute Atomic Mass

One of the more difficult computations in chemistry is the calculation of the atomic mass. The atomic mass is an average mass of all the naturally occurring isotopes. The calculation of the mass uses fractions rather than the usual route we use to compute an average. Let's start by getting a feel for averaging.

Suppose you have three students who take a test, and their grades on the test are 71, 83, and 95. What is the average score of the three students?

This is a straightforward average problem. Just add the three scores together and divide by three:

Average score = (71 + 83 + 95)/3
= 83

Now imagine that you have six students—three students get a 71, one gets an 83, and two get 95's. What is the average score for these six students?

Average mass = ([3 × 71] + [1 × 83] + [2 × 95])/6
= 81

It is also possible to report the *fractional* number of students receiving each of the three scores. In the above case, we could say that half of the students scored a 71, 1/6 of the students scored 83's and 1/3 of the students scored 95's. We could even report percentages: 50% of the students scored 71's, 16.67% of the students scored 83's, and 33.33% of the students scored 95's.

To calculate averages from fractional or percentage data, we use:

$$\text{Average} = (f_1 \times \text{score}_1) + (f_2 \times \text{score}_2) + (f_3 \times \text{score}_3) + \ldots$$

Using the data from the above case:

$$\text{Average} = (0.50 \times 71) + (0.1667 \times 83) + (0.3333 \times 95) = 81$$

Compute the average for the following situations.

1. Twenty students took a 10-point quiz on atomic structure. The grades on the quiz were as follows: two students got 4's, three students got 5's, four students got 6's, one student got a 7, five students got 8's, three students got 9's, and one student got a 10. What was the average score on the quiz?
2. In a different class, the quiz on atomic structure produced the following results:

Score	Percent Receiving Score
0	0
1	0
2	0
3	0
4	10
5	15
6	25
7	20
8	15
9	10
10	5

What is the average score in this class? (**Hint:** Beware—the percent is reported, not the fraction. How do you convert a percent to a fraction?)

3. Next, put the same mathematical procedures in the context of chemistry. Just replace score with atomic mass of the isotope.

Isotope	Atomic Mass of the Isotope	Percent Abundance
24-Mg	23.98504	78.70
25-Mg	24.98584	10.13
26-Mg	25.98259	11.17

What is the average atomic mass of magnesium? Check your answer with your periodic table. Note that the word "average" is dropped from atomic mass. The atomic mass of an element is the average atomic mass of the individual isotopes of that element.

Compute the atomic mass of each of the following two elements.

1. Silicon

Isotope	Atomic Mass of the Isotope	Percent Abundance	Fractional Abundance
28-Si	27.97693	92.21	0.9221
29-Si	28.97649	4.70	0.0470
30-Si	29.97376	3.09	0.0309

2. Calcium (you fill in the fractional abundances).

Isotope	Atomic Mass of the Isotope	Percent Abundance	Fractional Abundance
40-Ca	39.9629	96.947	
42-Ca	41.95863	0.646	
43-Ca	42.95878	0.135	
44-Ca	43.95549	2.083	
46-Ca	45.9537	0.186	
48-Ca	47.9524	0.18	

Practice 3 Computing Percent Abundance for the Special Case of Two Naturally Occurring Isotopes

When there are only two naturally occurring isotopes, it is possible to compute the percent abundances when the atomic mass unit and the isotope masses are known. Consider chlorine, which has an amu of 35.453.

Isotope	Atomic Mass	Percent Abundance
35-Cl	34.96885	x
37-Cl	36.9474	$100 - x$

We can use the equation for computing averages to solve for x:

$$35.453 = (x \times 34.96885 + [100 - x] \times 36.9474)/100$$

Notice that we have to divide by 100 to convert the percent abundance to a fractional abundance. Solving this equation for x:

$$3545.3 = 34.96885x + 3694.74 - 36.9474x$$

$$149.44 = 1.97855x$$

$$x = 75.53$$

Filling in the table we find the percent abundances of the two chlorine isotopes to be:

Isotope	Atomic Mass	Percent Abundance
35-Cl	34.96885	75.53
37-Cl	36.9474	$100 - x = 24.47$

Compute the percent abundances for the following elements.

1) Copper, amu = 63.54

Isotope	Atomic Mass	Percent Abundance
63-Cu	62.9298	
65-Cu	64.9278	

2) Galium, amu = 69.72

Isotope	Atomic Mass	Percent Abundance
69-Ga	68.9257	
71-Ga	70.9249	

Practice 4 Balancing Nuclear Decay Reactions

Balancing nuclear decay reactions is similar to (and in many cases easier than) balancing chemical equations. We just have to realize that the total mass and the total charge cannot change in nuclear decay. It is possible to change a proton into a neutron but if we do that, then we must generate another particle with a positive charge. Let's look at the specific decay reactions:

1. Decay producing alpha particles

Consider 238 U $\rightarrow$ 4 He + 234 Th

The text asks you to check that this radioactive decay reaction balances. First, add up the protons. There are 92 protons on the uranium side. On the right-hand side, there are 2 protons from the alpha particle and 90 from thorium. 90 + 2 gives us 92 protons, also on the right-hand side.

The total mass number also remains the same on both sides. On the left side of the equation, the mass number is 238. On the right side, it is 4 + 234 = 238.

What does this say about the number of neutrons on both sides?

Fill in the table below for some radioactive isotopes that decay by emitting an alpha particle.

Radioactive isotope	Number of Protons	Number of Neutrons	Products	Balanced Equation
Gd-152	64	88	Alpha, Sm-148	
Au-185	79		Alpha,	
Hg-187			Alpha,	

2. Decay producing beta particles

Just like alpha emission, the total number of protons and the total mass number has to be conserved in beta emission. Therefore, we can predict the products of nuclear decay reactions that produce beta particles. Fill in the following table for some radioactive isotopes that decay by emitting a beta particle.

Radioactive isotope	Number of Protons	Number of Neutrons	Products	Balanced Equation
Tl-204	81	123	Beta, Pb-204	
Te-129	52		Beta,	
Ru-106			Beta,	

Practice 5 Predicting Reaction Products

As long as you know either the decay mode or the decay product you can accurately predict either the mode or product. Fill in the table below with the missing information.

Radioactive isotope	Products	Balanced Equation
Hf-181	Beta,	
Lu-176	?, Hf-176	
Pt-190	?, Os-186	

Practice 6 Using Half-lives to Compute Amount of Radioactive Material Remaining

Look at Table 12-5, which lists some half-lives of five isotopes. Use that data to answer the following questions:

a. How long are four half-lives for iron-59?

b. If you begin with a 10-g sample of iron-59, how much remains after four half-lives?

c. You note that after 180.4 days you have 3 grams of iron-59. How much iron-59 did you start with?

d. How many half-lives will it take to reduce 100 grams of I-131 to 1.5625 grams?

e. How long will it take to reduce 100 grams of I-131 to 1.5625 grams?

CHAPTER 13

Electrons and Chemical Bonding

Checklist

What you need to be able to do when you finish Chapter 13

- Predict the shape of a molecule from its Lewis structure.
- Use Figure 13-12 to relate the wavelength of light to its energy.
- Be able to sketch the shapes of the *s* and *p* orbitals.
- Using the periodic table, write the electron configurations of any atom in the periodic table. Pay particular attention to the atoms of the elements in the first three rows.
- Know the trends in the ionization energies of the elements. Be able to explain why the first ionization energies of the noble gases are so much higher than the first ionization energies of the Group 1 metals.

Practice Problems and Study Hints

Practice 1 Molecular Shapes from Electron Domains

There are three critical steps in the process of predicting molecular shape of a given molecule from its Lewis structure. First, you have to count electron domains correctly. Second, you have to arrange these electron domains in the correct geometric shape. Third, you have to ignore the lone pairs (if any) and "see" the resulting arrangement of the atoms.

1. The molecules methane (CH_4), ammonia (NH_3), and water (H_2O) clearly illustrate the process of determining the molecular geometry.
   a. Write the Lewis structure of all three molecules.
   b. How many electron domains does each molecule have?
   c. What is the geometry of the electron domains?
   d. What is the geometry of the molecule?

2. Sometimes it is tricky to see the geometry of the molecule, or the geometry around a particular atom. It is helpful if you can sketch the different geometries on your paper. Notice that for the electron domains listed below, there are only three geometries:

Electron Domain	Geometry
2	Line
3	Trigonal
4	Tetrahedral

   a. Sketch each of the geometries on your paper. Note that the tetrahedral is the most difficult because it is in three dimensions.
   b. Looking at your drawings, imagine that one of the domains is actually a lone pair. Cover up that domain with your hand or with a piece of paper. What is the geometry of the remaining pair?

3. Look at Table 13-1. Below are nine molecules. Match the molecules below with the appropriate entries in Table 13-1. (Write the Lewis structures for each of the molecules.)

   H_2CO (consider carbon), H_2CO (consider oxygen), CO_2 (consider carbon), HCN (consider carbon), HCN (consider nitrogen), HNO (consider nitrogen), CF_4, Cl_2O (oxygen is central atom), NH_3

   Try It Yourself: Can you predict the correct molecular shapes for the molecules in problems 4 and 5 on page 448?

Practice 2 Relating Energy to Wavelength of Light

The approximate wavelengths associated with color of visible light are listed in the table below.

Color	Wavelength (nm)
Violet	400
Blue	460
Green	500
Yellow	560
Orange	620
Red	700

Use the above table and Figure 13-12 in your textbook to answer the following questions:

a. Hydrogen gas emits an intense blue light when excited with electricity. What is the energy of the light if one mole of the hydrogen gas is emitting the light?
b. The electrical discharge of helium produces a bright yellow line. What is the energy of the light if *two* moles of helium gas are emitting the light?

Practice 3 Atomic Orbitals

1. Atomic orbitals are associated with definite shapes. You should be able to sketch the *s* and *p* orbitals. Refer to pages 453 and 454. Sketch the shape of the *s* and *p* orbitals so that you can see them more clearly in your mind.
2. Atomic orbitals are also associated with energy levels. These energy levels are sometimes called shells and subshells. Look at Table 13-3 and pay particular attention to the subshell designation. It is into these subshells that we will add electrons. To do this, we need to know how many electrons can go into each subshell. Fill in the table below with the appropriate numbers:

	s orbital	*p* orbital	*d* orbital	*f* orbital
Number of Electrons				

3. Notice that the periodic table is arranged according to the orbitals that hold the valence electrons. Look at the periodic table, and answer the following questions:
   a. Which groups have the valence electrons in the *s* orbitals?
   b. Which groups have the valence electrons in the *p* orbitals?
   c. Which groups have the valence electrons in the *d* orbitals?
   d. Which groups have the valence electrons in the *f* orbitals?

Practice 4 Electron Configurations

To write electron configurations we need to know 1) the total number of electrons in the atom, 2) how many electrons go in each subshell, and 3) the order of the subshells. Let's consider sodium as an example. Sodium has 11 electrons. We already know how many electrons go in each subshell (refer to the table in Practice 3 of this chapter). Thus, we need to know only the order of the subshells. The order is:

1s, 2s, 2p, 3s, 3p, 4s, 3d, 4p, 5s, 4d, 5p, 6s, 4f, 5d, 6p, 7s, 5f, 6d

a. Can you see any relationship between this order and the periodic table?
b. Use the order given to construct the electron configurations of all the elements in the first three rows.

Try It Yourself: What is the electron configuration of the sodium ion Na^+? (**Hint:** How many electrons does the sodium ion have?)

Practice 5 Ionization Energies

We can use the electron configurations of atoms to understand the observed trends in ionization energies. Look at problem 32 in your text (page 463).

a. What is the electronic configuration of just the valence electrons in Group 1 elements?
b. What trend do you observe in the ionization energies of the Group 1 elements?
c. Explain the trend in the ionization energies.
d. What trend do you observe in the ionization energies of the Group 18 elements?
e. Can you construct a general statement about ionization energies from these data?

CHAPTER 14

Solutions, Molarity, and Stoichiometry

Checklist

What you need to be able to do when you finish Chapter 14

- Compute the molarity of a solution from the mass of solute.
- Convert between molarity and volume.
- Compute the molarity of ions in a solution given the molarity of the salt.
- Use molarity to calculate moles of reactants or products in a chemical reaction.

Practice Problems and Study Hints

Practice 1 Review the Mole-to-Mole Conversions in Chemical Transformations

A central concept in chemistry is the mole-to-mole conversions of compounds as expressed in chemical reactions. You already have a lot of experience with these conversions when the amount of reactants or the products is measured in mass. We can diagram the conversions as follows:

Grams of Reactants → **Moles of Reactants** → **Moles of Products** → Grams of Products

For example, use the following balanced chemical equation to compute the mass of carbon dioxide produced in the reaction. In this example, you will start with 22 grams of octane:

$$2C_8H_{18} + 25O_2 \rightarrow 18H_2O + 16CO_2$$

As we solve this problem, let's reflect on the steps leading to the solution:

1. *Convert mass of reactants to moles.*

   The molar mass of octane is 114 g/mol. Therefore, setting up the proportion:

$$\frac{\text{1 mole of octane}}{\text{114 grams}} = \frac{x \text{ mole of octane}}{\text{22 grams}}$$

   Cross-multiplying gives: $22 = 114x$, therefore, $x = 0.19$ moles octane.

2. *Convert moles of reactants to moles of products.*

From the balanced chemical equation, we see that for every 2 moles of octane reacting we produce 16 moles of carbon dioxide. By setting up the proportion and solving for x, we get:

$$\frac{16 \text{ moles of carbon dioxide}}{2 \text{ moles of octane}} = \frac{x \text{ mole of carbon dioxide}}{2 \text{ moles of octane}}$$

Cross-multiplying gives us $3.09 = 2x$; therefore, $x = 1.55$ moles of carbon dioxide.

3. *Convert moles of products to grams of products.*

$x = 68.2$ grams of carbon dioxide

We can generalize this three-step process as follows:

$$\frac{1 \text{ mole of carbon dioxide}}{44 \text{ grams}} = \frac{1.55 \text{ moles}}{x \text{ grams}}$$

1. *Convert amount of reactant to moles. Use whichever conversion is most appropriate.*
2. *Convert moles of reactant to moles of product using the balanced chemical equation.*
3. *Convert moles of product to appropriate amount (mass) of product.*

If our substances are *pure* substances, then we will often use *grams* as the unit for the amount of material of the substance. However, an extremely important type of substance is a *solution,* which is a mixture of two (or more) pure substances. We will consider solutions containing a solute and a solvent (two components).

The natural measurement unit for solutions is volume, usually in milliliters (abbreviated mL). Thus, we need to determine how to convert the volume of a solution into moles. Before doing this, though, let's practice the central step in stoichiometry problems.

Try It Yourself: Compute the moles of all the products given the below information about the reactants for the following reaction.

$8FeO + S_8 \rightarrow 8FeS + 4O_2$

1. 22.3 moles of iron (II) oxide and excess sulfur
2. 104 kg of sulfur and excess iron (II) oxide

Practice 2 Computing Molarity

As the name suggests, there are two substances in a binary solution: the solute and the solvent. We generally designate the substance that is present in the lesser amount as the "solute" and the substance that is present in excess is termed the "solvent." Adding salt to water, for example, creates a solution. In this example, the water is the solvent and the salt, when dissolved in the water, is the solute. Water is so often the solvent that we can write the solute as NaCl(*aq*), where the "*aq*" denotes water. Here it is clear that the NaCl is the solute.

Say we were to prepare a solution of saltwater by dissolving 10 grams of salt into a given amount of water. We want to prepare 500 mL of this solution, therefore we will use 500 mL of water. The question then becomes, how do we describe the concentration of the salt in the water?

Clearly, one way to do it would be to label the flask 10 grams of NaCl per 500 mL of water (solvent). While this would work, it is not the traditional way to designate the concentration of solutions. Rather, it is more common to describe the concentration in terms of moles of solute per liter of solution.

How can we convert the 10 grams of NaCl/500 mL of solution to moles of solute/liter of solution? Let's work through the problem.

1. Convert the grams of solute to moles of solute. (I get 0.17 moles of NaCl.)
2. Convert the volume of the solution (milliliters) to liters. (I get 0.500 L.)
3. Divide the moles of solute by the volume of solution. (0.17/0.500 = 0.34.)
4. Label the solution 0.34 M (molar).

Practice this type of conversion by computing the molarity of the following solutions:

1. 25 grams of calcium chloride dissolved in water to make 1200 mL of solution.
2. 10 grams of glucose ($C_6H_{12}O_6$) dissolved in water to make 600 mL of solution.
3. 1.5 mg of sodium bromide dissolved in water to make 150 mL of solution.

Practice 3 Using the Molarity Equation in Other Ways

There are other problems we can solve using the molarity equation (M = moles solute/volume of solution).

1. *Preparing solutions of known molarity.* Chemists often need to prepare a solution of known molarity. For example, suppose I want to prepare 500 mL of a 1.2-M sodium chloride solution. How would I prepare this solution?
   a. First, I would compute the number of moles of sodium chloride needed to create a solution with a molarity of 1.2 M. Using the molarity equation:

      moles NaCl = MV
      moles NaCl = 1.2 moles/L × 0.500 L
      = 0.6 moles of NaCl are needed for a 500-mL solution

   b. Compute the number of grams needed to get the moles. To do this, I multiply the moles by the molar mass of the compound.

      grams NaCl = 0.6 moles × 58.5 grams/mole
      = 35.1 grams

   c. Now that I know how much NaCl I need to create the solution, I am ready to prepare it. To start, I weigh out 35.1 grams of NaCl and put it into a 500-mL flask. To finish the solution, I add water until the 500-mL mark is reached.

   Try It Yourself: Practice by writing out exactly what you would do to prepare the following solutions.

   a. 250 mL of 0.5 M solution of potassium nitrate.
   b. 1250 mL of 2.0 M solution of sodium phosphate (Na_3PO_4).
   c. 500 mL of 1.5 M solution of potassium hydroxide.

2) *Computing volumes of solutions.* This is the primary skill needed to solve stoichiometry problems. In this case, we may already know the molarity of the solution and the number of moles we want. Although, there are times when we will have to compute the number of moles desired for the solution. Once that number is known, however, the question becomes what volume do we take.

Rearranging the molarity equation, we can solve for V = moles/M.

Suppose, for example, that I want to know how many mL of a 1.2-M solution of KNO_3 I need to get 0.01 moles of the KNO_3.

$$\text{Volume} = \frac{0.01 \text{ moles}}{1.2 \text{ moles/liter}}$$

$$= 0.00833 \text{ liter}$$

$$= 8.33 \text{ mL}$$

Try It Yourself: Practice solving for volume by computing the volume required in order to obtain the desired moles of solute in each of the following problems.

a. 0.5 moles of NaCl from a 2.0-M stock solution

b. 1.2 moles of KCl from a 1.0-M stock solution

c. 0.75 moles of HCl from a 2.0-M stock solution

Practice 4 Seeing and Computing What Is in Solution

Aqueous solutions of common table salt (NaCl) or of sugar ($C_{12}H_{24}O_{12}$) are both binary solutions. The solute in the first case is the salt and in the second case is the sugar. However, each solute's behavior in the solution is quite different.

Sodium chloride is an ionic compound and thus, when dissolved in water, it dissociates into its constituent ions. Therefore, there is no "molecule" of NaCl present in the solution. Rather, there are individual sodium Na^+ and chloride Cl^- ions contained within the solution. On the other hand, sugar is a covalent compound and thus, when it dissolves in water, there are individual molecules of sugar present in the solution.

A 1.0-M aqueous solution in which the solute is a covalent compound can be thought of as containing one mole of the solute per liter of solution. Alternatively, a 1.0-M aqueous solution of an ionic compound has no moles of the "molecule." The actual molarity of the ions depends upon the ionic compound's formula unit.

For example, a 1.0-M solution of NaCl contains 1.0 mole of sodium ions and 1.0 moles of chloride ions in 1 liter of solution. Thus, the concentration of sodium ions is 1.0 M as is the concentration of the chloride ions. In contrast, a 1.0 M solution of $CaCl_2$ contains one mole of calcium per liter of solution but 2 moles of chloride. Thus, the concentration of calcium in the solution is 1.0 M, and the concentration of chloride is 2.0 M.

Try It Yourself: For each of the ionic compounds below, compute the actual molarity of the ions present in the solution.

a. 1.6-M solution of potassium chlorate ($KClO_3$).

b. 0.25-M solution of sodium hydroxide.

c. 1.5-M solution of iron (III) nitrate.

Practice 5 Solution Stoichiometry

We are now ready to go back to the original problem—how to compute products from reactants when the reaction occurs in solution. Let's modify the original model to include the solution state:

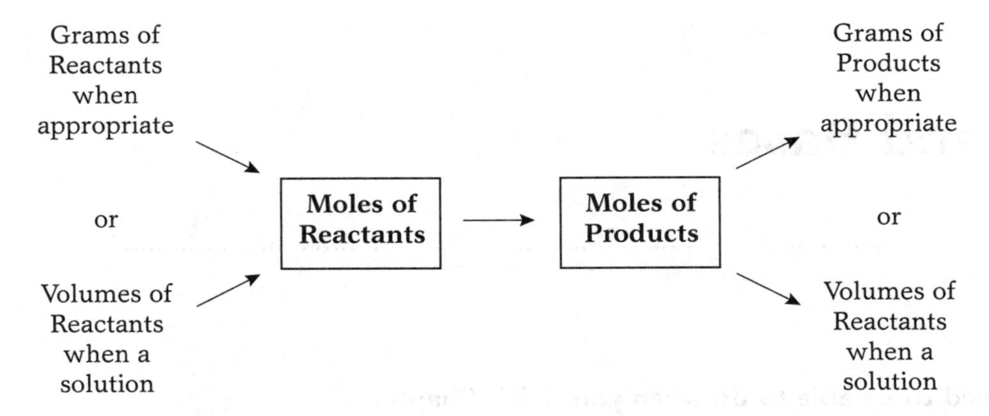

An example of a reaction in solution is the precipitation of silver chloride shown in the equation:

$AgNO_3\ (aq)\ +\ HCl\ (aq) \rightarrow AgCl(s)\ +\ HNO_3(aq)$

If I mix 50 mL of 1.0-M silver nitrate with 50 mL of 1.0-M hydrochloric acid, then what mass (in grams) of silver chloride do I produce?

1. *Compute the moles of reactants.*
   a. $AgNO_3$ Moles = 1.0 M × 0.050 L = 0.050 moles
   b. HCl Moles = 0.050 moles
2. *Compute the limiting reactant, if necessary.*
   In this case, both reactants are used up completely.
3. *Compute the moles of product (AgCl).*
   Because the reaction occurs in a 1:1 ratio, we must produce 0.050 moles of AgCl.
4. *Compute the mass of silver chloride from the moles and molar mass.*
   mass of AgCl = 0.050 moles × 143 g/mol = 7.2 grams

Try It Yourself: Practice this calculation by completing the following solution stoichiometry problems.

Iron (II) nitrate reacts with sodium phosphate to produce the precipitate iron (II) phosphate. The chemical reaction is:

$3Fe(NO_3)_2(aq)\ +\ 2Na_3PO_4(aq) \rightarrow Fe_3(PO_4)_2(s)\ +\ 6NaNO_3(aq)$

a. If I mix 50 mL of 1.0-M iron (II) nitrate with 100 mL of 2.0-M sodium phosphate, how many grams of iron (II) phosphate precipitate? Be sure to compute the limiting reactant.
b. If I mix 100 mL of 1.0-M iron (II) nitrate with 100 mL of 1.0-M sodium phosphate, how many grams of iron (II) phosphate precipitate?
c. You have mixed 100 mL of 1.0-M iron (II) nitrate with 50 mL of 2.0-M sodium phosphate. How many grams of iron (II) phosphate will precipitate? What species are left in your solution? What is the molarity of each of the species in the solution that you have created?

CHAPTER 15

Acids and Bases

Checklist

What you need to be able to do when you finish Chapter 15

- Distinguish between the three models of acids and bases: Arrhenius, Lewis, and Brønsted.
- Deduce the Lewis acid or base behavior from the Lewis structure.
- Identify conjugate acids and bases in the Brønsted model.
- Be able to compute the concentration of all species in a solution of a strong acid or strong base.
- Be able to calculate pH from the hydronium ion concentration as well as be able to calculate the hydronium ion concentration from the pH.
- Be able to calculate pOH from the hydroxide ion concentration as well as be able to calculate the hydroxide ion concentration from the pOH.
- Be able to calculate the pH from the pOH or the pOH from the pH.

Practice Problems and Study Hints

Practice 1 Models of Acids and Bases

There are three major models of acids and bases. The first model was developed by Arrhenius in 1883 and this model has been refined over the years. In 1923 Brønsted and Lowry published their model of acids and bases, which expanded upon the model proposed by Arrhenius. Lastly, in 1923, Lewis presented his initial ideas on an acid-base theory that was more inclusive than the Brønsted-Lowry model. Looking at the historical record of the three models shows us how each model is simply a more generalized version of the previous model. Let's examine these three models in their historical contexts.

In 1883, the nature of the atom—that is, the electrons and the nucleus—was postulated, but unproven. It is in this context that Arrhenius studied the behavior of solutions when electrical current was passed through them. From these studies he deduced that charged species must be in the water and thus deduced the existence of ions. Shortly after that, Arrhenius proposed that acids and bases were substances that produced ions in solution.

Notice again the theme in chemistry that we must relate the *structure* of the atom or molecule to its *behavior.* A unique behavior of acid and base molecules is their ability to conduct electricity. Pure water alone will not conduct electricity. The *explanation* for this behavior is the existence of ions; therefore, acids and bases must produce ions in the solution if they can conduct electricity. Thus, the Arrhenius model of an acid is something that produces hydrogen ions, and the Arrhenius model of a base is something that produces hydroxide ions.

Arrhenius is credited with defining acids and bases. Brønsted, Lowry, and Lewis were interested in a different unique behavior of acids and bases; namely, the fact that acids and bases readily react with each other. Thus, we now define acids and bases in terms of how they affect one another.

In the Brønsted-Lowry model shown below, a proton from the acid is transferred to the base.

$$H\text{-}Cl + OH^- \rightarrow H_2O + Cl^-$$

The HCl is the acid (proton donor) and the OH^- is the base (proton acceptor).

We can visualize the same reaction using a Lewis model. To do this, we need to draw all the electron dots:

$$H{:}\ddot{\underset{\cdot\cdot}{Cl}}{:} + {:}\ddot{\underset{\cdot\cdot}{O}}{:}H^- \rightarrow {:}\overset{H}{\underset{\cdot\cdot}{O}}{:}H + {:}\ddot{\underset{\cdot\cdot}{Cl}}{:}^-$$

The HCl (really the H^+) is the Lewis acid because it *accepts* an electron pair, and the OH^- is the Lewis base because it *donates* an electron pair.

You may have noticed that HCl is an Arrhenius, a Brønsted-Lowry, and a Lewis acid. Similarly, OH^- is also an Arrhenius, a Brønsted-Lowry, and a Lewis base. The fact that later theories have to include earlier ones is a common phenomenon in science. By adding on to earlier theories, later theories allow us to be able to generalize molecular behaviors or properties, as is the case with acid-base models:

Every Arrhenius acid or base is also a Brønsted-Lowry and Lewis acid or base.

Every Brønsted-Lowry acid or base is also a Lewis acid or base.

But be aware that the converse is not true! There are Lewis acids and bases that are *not* Arrhenius acids and bases. There are Lewis acids and bases that are *not* Brønsted-Lowry acids and bases. And there are Brønsted-Lowry acids and bases that are *not* Arrhenius acids and bases. We can picture this as a hierarchy from the most general to the more specific:

Lewis Acids or Bases
(Electron pair donors and acceptors)

↓

Brønsted-Lowry Acids or Bases
(Proton donors and acceptors)

↓

Arrhenius Acids and Bases
(Produce hydronium and hydroxide in aqueous solution)

Having this historical context, let's work with these different models.

The Arrhenius Model

This model is often applied simultaneously to both acids and bases. The basic premise of the Arrhenius model is that acids and bases produce ions in an aqueous environment. Table 15-3 lists the most common strong acids. All of these acids are Arrhenius acids.

Let's examine what these acids do in water.

Acids

All strong acids dissociate in water according to:

$$HNO_3 + H_2O \rightarrow H_3O^+ + NO_3^-$$

Try It Yourself: Write the equation for the dissociation of all the strong acids (from Table 15-3) in water.

Bases

Arrhenius bases are *hydroxides.* Hydroxides are composed of a metal ion with the hydroxide ion. For example, the water-soluble hydroxides are KOH, NaOH, CsOH, and RbOH—which are the Group 1 metal hydroxides. Like strong acids, water-soluble hydroxides dissociate completely in water as shown by the equation:

$$KOH + H_2O \rightarrow OH^-(aq) + K^+(aq)$$

The "(*aq*)" notation means "in water" (i.e., aqueous).

Try It Yourself: Write the equation for the dissociation of all the Group 1 hydroxides in water.

The Brønsted-Lowry Model

The Brønsted-Lowry Model focuses on the acid-base reaction in order to describe the behavior of acids and bases. The dissociation of an acid in water is viewed as an acid-base reaction. Consider the dissociation of nitric acid:

$$HNO_3 + H_2O \rightarrow H_3O^+ + NO_3^-$$

We can view this is as the dissociation of the nitric acid (as Arrhenius did) or we can view the nitric acid as a proton donor and the H_2O as the proton acceptor. We can even go one step further and label the products: the H_3O^+ becomes the *conjugate acid* and the NO_3- is the *conjugate base*. To see this, follow the proton:

$$\mathbf{HNO_3} + H_2O \rightarrow \mathbf{H_3O^+} + NO_3^-$$

The Brønsted-Lowry acid *donates* the proton leaving behind the conjugate base (the NO_3^-). The Brønsted-Lowry base *accepts* the proton and becomes the conjugate acid (the H_3O^+).

Try It Yourself:

1. For each of the Brønsted-Lowry acid-base reactions below, identify the acid, the base, the conjugate acid, and the conjugate base. Diagram, with a curved arrow, the proton transfer.
   a. $CH_3COOH + H_2O \rightarrow CH_3COO^- + H_3O^+$
   b. $NH_4^+ + H_2O \rightarrow NH_3 + H_3O^+$

c. $CH_3COOH + NH_3 \rightarrow CH_3COO^- + NH_4^+$
d. $HCl + NaOH \rightarrow H_2O + NaCl$

2. Make a list of the Brønsted-Lowry acids and bases from the above four reactions. Which ones are also Arrhenius acids and/or bases?

The Lewis Model

Similar to the Brønsted-Lowry model, the Lewis model is also based on the acid-base reaction. However, in the Lewis model, the focus is on the electron pair rather than the proton. The acid is the electron pair *acceptor* and the base is the electron pair *donor.*

Let's look at the reaction of HCl with OH^- from the Lewis model perspective. The proton on the HCl accepts an electron pair from the OH^- to form the chloride and the water. Because the HCl proton accepts the electron pair donated from the OH^-, it is the Lewis acid while the OH^-, the electron pair donor, is the Lewis base. The identification of the acid is sometimes confusing. It is usually easier to identify the base (the electron pair donor) first; that way, the acid (the other species) is apparent.

$$H{:}\ddot{\underset{\cdot\cdot}{Cl}}{:} + {:}\ddot{\underset{\cdot\cdot}{O}}{:}H^- \rightarrow {:}\overset{H}{\ddot{\underset{\cdot\cdot}{O}}}{:}H + {:}\ddot{\underset{\cdot\cdot}{Cl}}{:}^-$$

In this case, the HCl is also a Brønsted-Lowry acid and the OH^- is a Brønsted-Lowry base. In cases where there is no proton, the acid-base reaction is a Lewis acid-base reaction, not a Brønsted-Lowry. Recall that all Brønsted-Lowry acids are Lewis acids and all Brønsted-Lowry bases are Lewis bases. But this is a one-way relationship. Not all Lewis acids and bases are Brønsted-Lowry acids and bases. Here is an example:

$$6\ {:}\overset{H}{\ddot{\underset{\cdot\cdot}{O}}}{:}H + Al^{3+} \rightarrow Al\left[{:}\overset{H}{\ddot{\underset{\cdot\cdot}{O}}}{:}H\right]_6^{3+}$$

In this case, the aluminum ion is the electron pair *acceptor* and is therefore the acid (acceptor = acid). The HOH is the electron pair *donor* and is therefore the base.

Below are some more Lewis acid-base reactions. Identify the acid and the base:

a.

$$H{-}\underset{H}{\overset{H}{C}}{-}\underset{N}{\ddot{N}}{-}N + H^+ \rightarrow \left[H{-}\underset{H}{\overset{H}{C}}{-}\underset{N}{\overset{H}{N}}{-}N\right]^+$$

b.

$$H{-}\underset{H}{\overset{H}{C}}{-}\overset{O}{\overset{\|}{C}}{-}\ddot{\underset{\cdot\cdot}{O}}{:}^- + H^+ \rightarrow H{-}\underset{H}{\overset{H}{C}}{-}\overset{O}{\overset{\|}{C}}{-}\ddot{\underset{\cdot\cdot}{O}}{-}H$$

Practice 2 Calculating Concentrations

The only concentrations we will calculate are for strong acids and bases. Recall that strong acids and bases completely dissociate in water.

1. *Mole* calculations

   Consider the dissociation of HBr in water:

   $$HBr + H_2O \rightarrow H_3O^+ + Br^-$$

   How many moles of hydronium ion and bromide ion are in solution if:

   a. 1.0 moles of HBr are dissolved?

   b. 0.003 moles of HBr are dissolved?

   c. 2.5 moles of HBr are dissolved?

2) *Molarity* calculations

   Consider the same dissociation of HBr. Let's calculate the concentration of all the species in solution. To do this, we first have to determine what is in solution and then determine how much.

   i. *Determining what is in solution.* For each problem below, write the chemical equation for the dissolution of the acid or base in water.

   ii. *Determining the concentration.* For each species, compute the number of moles of the species in solution and then compute the molarity of the species (moles/volume of solution).

   a. How many moles of hydronium ion and bromide ion are in solution if 2.0 moles of HBr are dissolved?

   b. What is the concentration of the hydronium ion and bromide ion if we make a 1-liter solution of the 2.0 moles of HBr?

   For each of the following, assume you prepare a 500-mL solution. Compute the concentration of all species in the solution.

   c. 0.5 moles of HCl

   d. 1.5 moles of nitric acid

   e. 2×10^{-5} moles of perchloric acid

   f. 25.0 grams of NaOH

   g. 6.7 mg of KOH

Practice 3 Computing pH

The calculation of the pH of the solution is easy once you have computed the concentration. You should also be sure to calibrate your answer: If you are calculating an acid, the pH should be less than 7; if it is a base, then it should be greater than 7.

1. For each of the following, compute the pH of the solution:

   a. 1.0 M HCl

   b. 0.02 M HNO_3

   c. 2.5 M $HClO_4$

2. For each of the following, compute the pOH of the solution:
   a. 1.0 M NaOH
   b. 0.06 M KOH
   c. 0.50 M CsOH
3. For each of the following, compute the pH of the solution from the concentration and then compute pOH from pH + pOH = 14.
   a. 0.5 M HCl
   b. 0.065 M $HClO_4$
   c. 1.25 M HBr
4. For each of the following, compute the pOH of the solution from the concentration and then compute the pH from pH + pOH = 14.
   a. 2.0 M NaOH
   b. 1×10^{-3} M KOH
   c. 0.008 M CsOH
5. For each of the following compute both the pH and the pOH of the solution. First, decide which calculation is appropriate to start with.
   a. 0.8 M HCl
   b. 1.5 M NaOH
   c. 2.5×10^3 M HNO_3
   d. 0.055 M KOH
6. For each of the solutions below, compute the concentration of the hydronium ion or the hydroxide ion and then compute the pH or the pOH. Next, calculate the pH value for the basic solutions.
   a. 5.0 g of NaOH is dissolved in water to make 250 mL of solution.
   b. 10.0 g of HCl is dissolved in water to make 500 mL of solution.
   c. 25.0 g of KOH is dissolved in water to make 2 liters of solution.
7. Let's account for temperature in your calculations. This one is a bit tricky! Work problem 48 in your textbook. If the solutions in (6) above were prepared at 40 °C, what would be the pH and pOH of each of the solutions?

CHAPTER 16

Equilibrium Systems

Checklist

What you need to be able to do when you finish Chapter 16

- Write equilibrium expressions for equilibrium reactions involving concentrations.
- Write equilibrium expressions for reactions involving gases.
- Write equilibrium expressions for reactions that include solids and liquids.
- Calculate equilibrium constants from concentrations.
- Calculate equilibrium concentrations of metal-ligand complexes.
- Calculate pH of weak acid/base equilibrium.
- Calculate solubility of slightly soluble salts in water.

Practice Problems and Study Hints

Practice 1 Writing Equilibrium Expressions

Before you can work any equilibrium problems, you must be adept at writing equilibrium expressions. To write an equilibrium expression, you *must* start with a balanced chemical equation.

1. For example, consider the gas phase equilibrium:

$$N_2(g) + 3H_2(g) \rightleftharpoons 2NH_3(g)$$

Write the equilibrium expression for this gas phase reaction.

2. Next, consider the reaction for the decomposition of ammonia:

$$2NH_3(g) \rightarrow N_2(g) + 3H_2(g)$$

Write the equilibrium expression for this gas phase reaction.

3. Write one or two sentences describing the relationship between the two equilibrium expressions.

I hope it is apparent that we must have the chemical equation to write an equilibrium expression. Let's practice writing equilibrium expressions using the following rules:

i. If gases are present, then we include the partial pressure of the gas in the equilibrium expression.

ii. If solutions are present in the chemical equation, then we include the concentration of the species in the equilibrium expression.

iii. If a solid is present in the chemical equation, then we exclude the solid from the equilibrium expression.

iv. If a pure liquid is present in the chemical equation, then we exclude the liquid from the equilibrium expression.

4. Write equilibrium expressions for the following equilibria reactions:

a. $HF(aq) + H_2O(l) \rightleftharpoons H_3O^+(aq) + F^-(aq)$

b. $3Ca(NO_3)_2(aq) + 2Na_3PO_4(aq) \rightleftharpoons Ca_3(PO_4)_2(s) + 6NaNO_3(aq)$

c. $O_2(g) + O(g) \rightleftharpoons O_3(g)$

Practice 2 Calculating Equilibrium Constants

Our goal is to obtain a value for the equilibrium constant. To do so, we need two things: 1) the equilibrium expression, and 2) the concentrations of all the species in the equilibrium expression at equilibrium. For the three equilibrium expressions you wrote in Practice 1 above, compute the equilibrium constant for the following equilibrium situations:

a. $[HF] = 2.0$ M; $[H_3O^+] = 0.14$ M; $[F^-] = 0.014$ M

b. $[Ca(NO_3)_2] = 0.005$ M; $[Na_3PO_4] = 0.005$ M; $[NaNO_3] = 1.5$ M

c. $P_{O_2} = 1$ atm; $P_O = 0.001$ atm; $P_{O_3} = 1 \times 10^{-9}$ atm

Try It Yourself: Is your system at equilibrium? Consider the calcium phosphate precipitation equilibria (question (b) above). Is the following system at equilibrium?

$[Ca(NO_3)_2] = 0.001$ M; $[Na_3PO_4] = 0.001$ M; $[NaNO_3] = 1.0$ M

Practice 3 Metal-Ligand Equilibria, an Application of Equilibria

We are interested in computing the concentrations of species at equilibrium. This will require three steps: 1) we need a method for organizing all of the information in a table; 2) we need to know how to solve for the roots of a quadratic equation; and 3) we need to interpret the roots of our quadratic equation. We will do this process three times—for metal-ligand equilibria, for acid-base equilibria, and for solubility equilibria.

Consider the example in your text:

$$Zn^{2+}(aq) + C_3H_5O_3^-(aq) \rightleftharpoons Zn(C_3H_5O_3)^+(aq)$$

- First, which is the metal and which is the ligand?
- Look at problem 22 and identify the metal and the ligand.

For any given metal-ligand equilibria, we can be asked to complete one of two calculations: 1) given all equilibrium concentrations, what is the value of K? or 2) given K and the concentration of the metal-ligand complex, what is the concentration of the metal and the ligand?

In both cases, we must set up the equilibrium expression.

- Write the equilibrium expression for the following metal-ligand complexes:

$$Ag^+(aq) + NH_3(aq) \rightleftharpoons Ag(NH_3)^+(aq)$$
$$Cr^{3+}(aq) + CN^-(aq) \rightleftharpoons Cr(CN)^{2+}$$

Solving for K

Use your equilibrium expressions above to solve for the value of the equilibrium constant if:

- $[Ag^+] = 0.025$ M, $[NH_3] = 0.30$ M, $[Ag(NH_3)^+(aq)] = 0.5$ M
- $[Cr^{3+}] = 2.5 \times 10^{-6}$ M, $[CN^-] = 0.10$ M, $[Cr(CN)^{2+}] = 0.1$ M

Solving for Concentrations

Whenever we solve for concentrations, we can use a general method that involves making a table. (The "ICE" Table, see Figure 16-8.)

1. Write the equilibrium expression—in this case the metal-ligand equilbrium.
2. Write down (find) the equilibrium constant (K) that goes with the equilibrium expression.
3. Record the initial concentrations of the all the species.
4. Use the stoichiometry to determine the change in all the species. Notice that for acid-base equilibrium the stoichiometry is always 1:1:1.
5. Write down the mathematical expression for the concentration at equilibrium. Look at the last line in Figure 16-8. Notice that the equilibrium concentrations are written in terms of the variable x.
6. Once you have the table, use the table to set up the equilibrium expression, as is completed in your text.
7. Rearrange your expression in the form needed for the quadratic formula.
8. Solve the equation using the quadratic formula.
9. The solution is usually in terms of x. Refer back to your table to relate x to the answer desired. Always keep in mind what the question has asked you to find. Are you being asked to calculate the concentration of a product or a reactant? Write your answer clearly.

Practice computing equilibrium concentrations using the silver/ammonia metal-ligand equilibrium:

$$Ag^+(aq) + NH_3(aq) \rightleftharpoons Ag(NH_3)^+(aq)$$

Calculate $[Ag(NH_3)^+]$ if our initial concentrations if silver ion and ammonia are each 0.2 M.

To solve this problem, set up the ICE table. Fill in the final concentrations.

Balanced Equation:	$Ag^+(aq)$ +	$NH_3(aq)$ $\rightleftharpoons$	$Ag(NH_3)^+(aq)$
Initial Concentrations	0.2	0.2	0
Change	$-x$	$-x$	$+x$
Final Concentrations			

To solve for x, set up the equilibrium expression, rearrange, and solve the quadratic equation for x.

$$K = 66.7 = \frac{x}{(0.2 - x)(0.2 - x)}$$

$$66.7 = \frac{x}{0.04 - 0.4x + x^2}$$

$$66.7(0.04 - 0.4x + x^2) = x$$

$$2.668 - 26.68x + 66.7x^2 = x$$

$$66.7x^2 - 27.78x + 2.668 = 0$$

Notice there are several steps until we are ready to solve the quadratic equation. Just do each step slowly and accurately—it looks like a lot of steps, but if you do each one carefully you will get to the right answer.

Solving the quadratic equation gives:

$$x = \frac{27.78 \pm \sqrt{27.78^2 - 4(66.7)(2.668)}}{2(66.7)}$$

$$x = \frac{27.78 \pm \sqrt{59.91}}{133.4}$$

$$x = \frac{27.78 \pm 7.74}{133.4}$$

$$x = 0.266, 0.150$$

Whenever you solve the quadratic equation you will get two roots. Based on the situation that you are given, you now have to decide which answer is correct.

To decide, we need to think about the meaning of x. Look at your ICE table. Which value of x do you think is correct?

From the ICE table, one interpretation of x is the concentration of silver ion that complexes. This value cannot be larger than the amount of silver ion we start with. Therefore, we have to reject 0.266, and the correct value of x = 0.150M.

The question may have asked us to report the actual concentrations of all species present in solution. These are now:

$[Ag^+] = 0.05$ M; $[NH_3]$=0.05 M; $[Ag(NH_3)^+] = 0.150$ M

Try It Yourself: Practice the procedure for the silver/ammonia equilibrium. Compute the concentration of all species present in the solution if you start with the following combinations:

a. $[Ag^+] = 0.50$ M; $[NH_3] = 0.50$ M; $[Ag(NH_3)^+] = 0.0$ M
b. $[Ag^+] = 0.05$ M; $[NH_3] = 0.05$ M; $[Ag(NH_3)^+] = 0.0$ M
c. $[Ag^+] = 0.0$ M; $[NH_3] = 0.0$ M; $[Ag(NH_3)^+] = 1.0$ M
d. $[Ag^+] = 0.0$ M; $[NH_3] = 0.0$ M; $[Ag(NH_3)^+] = 0.2$ M

Practice 4 Calculating pH of Weak Acids

The calculation of pH of weak acids is much easier if you use the equilibrium table (the "ICE" table, see Figure 16-10). Again, let's review the logic for the construction of the table:

1. Write the equilibrium expression—that is, the dissociation of the weak acid.
2. Find the equilibrium constant (K_a) that goes with the equilibrium expression.
3. Record the initial concentrations of all the species.
4. Use the stoichiometry to determine the change in all the species. Notice that for acid-base equilibrium the stoichiometry is always 1:1:1.
5. Write down the mathematical expression for the concentration at equilibrium. Look at the last line in Figure 16-10. Notice that the equilibrium concentrations are written in terms of the variable x.
6. Once you have the table, use the table to set up the equilibrium expression, as it is completed in your text.
7. Rearrange your expression in the form needed for the quadratic formula.
8. Solve the equation using the quadratic formula.
9. The solution is usually in terms of x. Refer back to your table to relate x to the answer desired. Be certain that you are answering the question being asked!

Try It Yourself: Look at Table 16-2. Pick an acid (any acid) and calculate the pH of a 1.0 M solution of that acid. Repeat the process for at least two other acids.

Approximations Involving pH Calculations

Sometimes, you can shorten the quadratic equation step by using an approximation method. Suppose the ionization of the acid is very small. Then we can approximate the equilibrium concentration of the acid as equal to the initial concentration. Let's rewrite Table 16-10 making this approximation and assuming we begin with 1 M formic acid:

Balanced Equation:	$HCOOH(aq)$ +	$H_2O(l)$ $\rightleftharpoons$	$H_3O^+(aq)$ +	$HCOO^-(aq)$
Initial Concentrations	1.0	—	0	0
Change	$-x$	—	x	x
Final Concentrations	1.0	—	x	x

Notice that the "true" equilibrium concentration of the HCOOH is $1.0 - x$. However, we are approximating the equilibrium concentration of the acid to be the same as the initial concentration.

Using this table, set up the equilibrium expression:

$$1.77 \times 10^{-4} = \frac{x^2}{1}$$

This is a dramatically simpler mathematical expression because we do not need to use the quadratic equation:

$$x^2 = 1.77 \times 10^{-4}$$

$$x = \sqrt{1.77 \times 10^{-4}}$$

$$x = 0.013$$

Try It Yourself: Redo the calculations you performed earlier and calculate the pH of your three acids making the approximation above. Does this method change your pH?

There is always a question concerning when you can use the approximate method and when you have to use the quadratic equation. There are two answers to this. The chemical answer is: You can use the approximation method when the acid does not dissociate much. This occurs when you have either a very small K_a ($\sim 10^{-7}$) or a fairly large initial concentration (1 M).

Mathematically, you can use the approximation by computing the estimated percentage of dissociation and comparing that number to 10.

In the formic acid example above, we obtain a value of $x = 0.013$. The percentage of dissociation of the acid is:

$$\frac{x}{10} \times 100 = \frac{0.013}{1.0} \times 100 = 1.3\%$$

Because 1.3% is less than 10%, we can conclude that the acid is very weakly dissociated and the approximate method is acceptable.

Try It Yourself: For the three acids above, compute the approximate percentage of dissociation. Is it less than 10%? Does this agree with your pH calculations?

Practice 5 Solubility, Another Application of Equilibria

Once again we need to modify our prior understanding of solubility. Salts are no longer soluble or insoluble—they can also be slightly soluble. In fact, we can imagine writing an equilibrium equation for every salt:

$$\text{Salt }(s) + H_2O \rightleftharpoons \text{Individual ions }(aq)$$

For example:

$$FePO_4(s) + H_2O \rightleftharpoons Fe^{3+}(aq) + PO_4^{3-}(aq)$$

Or

$$NaCl(s) + H_2O \rightleftharpoons Na^+(aq) + Cl^-(aq)$$

If the equilibrium lies far to the right (the equilibrium constant is *large*), then we would say the salt is soluble. On the other hand, if the equilibrium lies far to the left (the equilibrium constant is *small*) then we would say the salt is insoluble. And if the equilibrium constant is in the middle, the salt is slightly soluble.

These categories are not hard and fast rules—you do not need to memorize the categories. But it is helpful to get a feel for the solubility based on the value of the equilibrium constant.

Look at the Table 16-3. Write the equilibrium equation, and the equilibrium expression for the most soluble and the least soluble salt in the table.

To calculate solubility, we can construct a table similar to the ICE table for acids and bases. Let's compute the solubility of silver iodide in water.

Balanced Equation:	$AgI(s)$ +	$H_2O(l)$ $\rightleftharpoons$	$Ag^+(aq)$ +	$I^-(aq)$
Initial Concentrations	—	—	0	0
Change	—	—	x	x
Final Concentrations	—	—	x	x

Notice that the only difference between this table and the table for acids and bases is the exclusion of the concentration of AgI. Why is it excluded?

Use the table to set up the equilibrium expression. In this case, the equilibrium constant is called K_{sp}.

$$K_{sp} = 8.5 \times 10^{-17} = x^2$$

$$x = \sqrt{8.5 \times 10^{-17}}$$

$$x = 9.2 \times 10^{-9}$$

It now remains to interpret x. Recall that in the acid/base application of equilibrium we often had to compute the pH = $-\log x$. In solubility equilibria, we will want to interpret x as the number of moles of salt that can be dissolved to make a 1-liter solution. Thus, the solubility of AgI is 9.2×10^{-9}.

Try It Yourself: Pick three salts (the most soluble, the least soluble, and one in the middle) from Table 16-3, and compute the solubility of these salts in water at 25 °C.

Once we can do these calculations, we can also compute the value of K_{sp} from solubility data. For example, we know that the solubility of iron(II) carbonate is 5.7×10^{-6}. What is the value of K_{sp}?

To solve this problem, we simply remember that x = solubility. Therefore, $x = 5.7 \times 10^{-6}$. And $K_{sp} = x^2 = 3.2 \times 10^{-11}$.

Try It Yourself: Consider the three salts for which you have computed the solubility. From *your* solubility calculations, compute the value of K_{sp}. Does your answer agree with the value in the table?

CHAPTER 17

Organic Chemistry and Biochemistry

Checklist

What you need to be able to do when you finish Chapter 17

- Know the common bonding patterns for carbon, fluorine, nitrogen, oxygen, and hydrogen.
- Sketch Lewis structures using these bonding patterns.
- Recognize and construct structural isomers of organic molecules.
- Distinguish saturated structures from unsaturated structures.
- Identify organic functional groups.
- Recognize the structures of biochemical molecules: carbohydrates, amino acids, and nucleic acids.

Practice Problems and Study Hints

Practice 1 Become Familiar with the Bonding Patterns of H, C, O, N, F
It is helpful to truly master the bonding patterns of these five atoms. Let's first reproduce Table 17-1 but add an additional column: the number of lone pairs. Fill in the table below. The oxygen row is filled in as an example.

Element	Number of Bonds	Electrons in Lone Pairs	Number of Lone Pairs
H			
C			
N			
O	2	4	2
F			

1. Look at lots of structures and be sure the bonding patterns above are followed. Look at the molecules in Example 17.5. Do each of these molecules fit the bonding patterns above?
2. Writing Lewis structures for organic molecules can be very cumbersome. Thus, we use shorthand notations to draw the structures. For example, we will often leave out the lone pairs on the nitrogen and oxygen atoms in the structure. You must be aware that the lone pairs are *assumed* to be there. Look at the structures in Figure 17-5. Do all of these structures follow the bonding rules above? Redraw each of the three structures to include the lone pairs on the oxygen.
3. Another shorthand notation removes the explicit bond between the C and H. Instead the group CH, CH_2, or CH_3 is written. Look at Figure 17-11. Do all of the structures follow the bonding rules? Look at the structures in Figure 17-22. Do all of the structures follow the bonding rules? Redraw the structures to show the explicit bonds between the C and H atoms. Include the lone pairs on the appropriate atoms. Do all of the atoms (except H) follow the octet rule?

Practice 2 Writing Lewis Structures for Organic Molecules

We can begin to construct the skeleton structure of organic molecules just knowing the bonding rules and the chemical formula. Sometimes, there are several structures possible, but, for small molecules, there is often only one possible structure.

1. Draw the skeleton structure, and then the full Lewis structure (include the lone pairs if required) for each of the following molecules. Use the H atom to help figure out the structure—remember H can form only one bond to an atom. Thus, H has to bond to a C, N, or O, but never to another H.

   a. CH_4 b. C_2H_6 c. C_2H_4

2. The following chemical formulas are not valid organic molecules. Try drawing a Lewis structure for each formula and explain why it is not valid:

   a CH_5 b. C_2H_5 c. C_2H_3

3. Next, include the O and N atoms. Draw valid Lewis structures for the following molecules. Do any of the following have more than one possible valid Lewis structure?

   a. CH_2O_2 b. C_3H_6O c. CH_5N

Practice 3 Structural Isomers

There are two valid Lewis structures for C_3H_6O. (Did you find both?) What is the same about the two structures and what is different?

Structural isomers are molecules that have the same chemical formula but different arrangements of atoms or different bonding patterns in the molecule. When we consider larger molecules structural isomers emerge.

Recognizing whether or not two molecules are isomers requires two steps. First, construct the chemical formula for the two molecules and compare. If the formulas are different, then the two molecules are not isomers. If the two molecules have the same chemical formula, then the two structures may be isomers. Now it gets tricky because we have to pay attention to the bonding in the molecules.

1. Look at the isomers of C_5H_{10} in Example 17-7. Consider only the arrangements of the five carbons in a row. Explain, in your own words, why there are only two isomers (not four) for this arrangement.
2. Now consider the formula C_5H_8. Consider, for now, only the arrangement of the carbon atoms in a row. How many isomers can you find? (I found four.)
3. Can you construct another isomer for C_5H_8 that has a ring structure? (There are many, try to find one.)

Practice 4 Recognizing Saturated Compounds

Saturated and unsaturated organic compounds play important roles in our nutrition. Saturated compounds have no double or triple bonds between two carbon atoms. While there might be a double bond between an oxygen atom and a carbon atom, if there is no carbon-carbon double or triple bond, the molecule is still saturated. Unsaturated compounds, on the other hand, have at least one multiple bond between two carbon atoms.

You cannot determine if a molecule is saturated or unsaturated unless you have the structural formula of the compound. Look at the molecules in **Practical "A"** in your text. Are there any unsaturated compounds? Look at the molecules in Figure 17-5. Are there any unsaturated compounds? (**Hint:** I found four unsaturated molecules. Do you agree?)

Practice 5 Recognizing Organic Functional Groups

Certain bonding patterns in organic molecules belong to particular categories. You should be familiar with some of these. Again, look at the molecules in **Practical "A."** These five groups are common organic groups.

Circle the bonding pattern in the molecular example that makes each molecule belong to the designated group. For example, you would circle the OH in the alcohol. In words, write down:

a. the definition of an ether

b. the definition of a carbonyl group

c. the definition of an acid group

d. the definition of an ester

Next, work through Example 17-11 and compare your answers.

Practice 6 Recognizing Common Biological Molecules

Biological molecules are often complex organic molecules containing at least one, and often more than one, functional group. There are three types of biological molecules you should be familiar with: carbohydrates, amino acids, and nucleic acids. You probably do not need to memorize particular compounds but you should know what distinguishes the three types of biological molecules. As you can guess the distinguishing features are the function groups.

Look at Figure 17-17. What functional groups do the two carbohydrates in (a) and (b) have? (I see two groups.) Look at the two carbohydrate molecules. What are the similarities between the two molecules? What are the differences?

Next, look at the amino acids listed in Table 17-3. What functional groups do you see in the amino acid? How is it different from the carbohydrate. Last, look at the nucleic acid structures in Figure 17-22. How do these differ from amino acids? From carbohydrates?

Make a table that lists the characteristic and distinguishing feature of each type of biological molecule. I filled in some information in the carbohydrate row.

Biological Type	Distinguishing Molecular Feature(s)	Chemical Formula of an Example Molecule
Carbohydrates	Contains OH groups. The number of OH groups seems to be one less than the number of carbon atoms.	$C_5H_{10}O_5$
Amino Acids		

Solutions to Odd-Numbered Text Exercises

CHAPTER 1

1.1 • Homogeneous means the mixture has the same properties throughout any given sample of it.

• Heterogeneous means the mixture has different properties in different samples of it.

1.3 Nitrogen, oxygen, water vapor, carbon dioxide

1.5 The components of a mixture can be separated by physical means (filtration, distillation, etc.), whereas a compound cannot be broken down physically.

1.7 a) An element is the simplest form of matter. An element contains only one kind of atom, e.g., hydrogen, nitrogen, oxygen, chlorine, sodium, iron.

b) A compound consists of two or more elements chemically bonded to each other in a particular ratio. This—definite composition and definite proportion—gives a compound the same properties throughout any sample of it.

1.9 A compound contains at least two elements. Therefore, examples a and c are elements and examples b and d are compounds.

1.11 a) Be b) O c) He

1.13 a) nitrogen, hydrogen, phosphorus, and oxygen.

b) calcium, carbon, hydrogen, oxygen

c) aluminum, chlorine, oxygen

1.15 The number of protons is equal to the atomic number. Looking at the periodic table, you can find the symbol that goes with each atomic number.

Atomic number	Number of protons	Symbol	Name
14	14	Si	silicon
94	94	Pu	plutonium

1.17 The Atomic number equals the number of protons. In a neutral atom, the number of protons equals the number of electrons and the mass number equals the number of protons plus the number of neutrons.

Element	Symbol	Number of protons	Number of neutrons	Number of electrons	Atomic number	Mass number
Lithium	Li	3	4	**3**	3	**7**
Cadmium	Cd	48	**65**	**48**	48	113
Iodine	I	53	74	**53**	**53**	**127**
Tungsten	**W**	74	**111**	**74**	74	185

1.19 See Web Animator on *The Practice of Chemistry* Web site.

1.21 A chemical formula gives the symbols of the elements it contains. Subscript numbers following the symbol give the ratio of elements to each other. If there is no numerical subscript after the symbol, the element occurs once in the formula.

1.23 a) Cannot determine from this information. This information tells us that the substance is a pure substance, but we cannot distinguish if this is either an element or a compound.

b) If oxygen is given off, that means the original substance contained oxygen. Hence, this is a compound.

c) If it cannot be separated into a simpler substance but it burns, this means that it is an element. If it were a compound, it wouldn't burn.

1.25
- Above –119.9°C, it is a gas.
- Between –119.9°C and –192°C, it is a liquid.
- Lower than –192°, it is a solid.

a) Gas b) Gas c) Gas d) Gas e) Liquid

1.27 Nonmetals generally are dull, brittle, good insulators of heat and electricity, and have low melting points and low boiling points. These are physical properties of nonmetals.

1.29 The formulas will have the general formula MN_2.
a) $CaCl_2$ b) BaI_2 c) MgF_2

1.31 a) True

b) False: There is only one semimetal in group VI and the rest are nonmetals.

c) False: All group VII are nonmetals.

1.33 a) Carbon is an element; sugar is a compound.

b) Hydrogen is an element; water is a compound.

c) Oxygen is an element; water is a compound.

1.35 Carbon, phosphorus, nitrogen, and sulfur are all nonmetals, hence, all the compounds are nonmetal oxides.

1.37 The product is a gas (physical property). Because it "burst into flame" we know that it combined with oxygen because burning is a rapid reaction with oxygen. This means that the product gas contains oxygen, so it can be decomposed to give oxygen back (chemical property).

1.39 If there is more than one element it is a compound. Therefore, a, e, f, h, and i are elements and b, c, d, g, and j are compounds.

1.41 a) This is a physical change that separates the board into two pieces.

b) This is a chemical change because the substance water (H_2O) is converted into different substances (H_2 and O_2).

c) This is a physical change of state in which a substance in the solid state is going directly to the gaseous state (sublimation).

d) This is neither a physical nor a chemical change. The tennis ball is not changed in any way by the act of bouncing.

e) This is a physical change of state (melting).

1.43 a) Cannot determine from this information. This information tells us that the substance is a pure substance, but we cannot distinguish if this is either an element or a compound. A mixture would melt at several different temperatures because each substance in the mixture would have a different melting point.

b) The gaseous substance is made of oxygen and something else, hence it is a compound.

c) If it cannot be separated into a simpler substance but it burns, this means that it is either an element or a compound. Either of these can burn to form a new substance.

1.45 The number of neutrons can be obtained by subtracting the atomic number from the atomic mass: 60 − 27 = 33 neutrons.

1.47 Cd has atomic number 48. The mass is 48 + 67 = 115. The atomic symbol is: $^{115}_{48}$Cd.

1.49

Action	Physical change?	Chemical change?
A solid dissolves in a liquid and produces heat.		The evolution of heat indicates a chemical reaction.
Bubbles of gas are formed when metal is added to a solution of acid.		The formation of gas bubbles indicates that there was a chemical change.
Copper metal can be drawn into copper wires.	This is a physical change. Only the shape is changed.	
Phosphorus burns when put into water.		Burning is a chemical change.

1.51 A compound contains at least two elements. Therefore, examples a and b are elements and c and d are compounds.

1.53 a) Na b) P c) F

1.55 The atomic number is equal to the number of protons. In a neutral atom, the number of protons equals the number of electrons, and the mass number equals the number of protons plus the number of neutrons.

Element	Number of protons	Number of neutrons	Number of electrons	Atomic number	Mass number	Atomic symbol
Chlorine	17	18	17	**17**	**35**	Cl
Barium	56	**81**	56	56	137	Ba
Radium	**88**	**138**	88	88	226	Ra

1.57

Symbol	Name	Period	Group
K	Potassium	4	I
N	Nitrogen	2	V
O	Oxygen	2	VI
Cl	Chlorine	3	VII
Ne	Neon	2	VIII
S	Sulfur	3	VI

1.59

Description	Physical change?	Chemical change?
Liquid ether can be	Freezing a liquid is a physical change of state.	
Candle wax melts.	Melting wax is a physical change of state.	
A solid is formed when two clear liquids are poured together.		Forming a solid from two liquids is a chemical change.
Sodium carbonate dissolves in acid, producing heat.		The sodium carbonate is reacting with the acid to produce heat, hence, a sign of a chemical change.
Hydrogen gas is formed when metal is added to a solution of acid.		The formation of a new substance (hydrogen gas) indicates that a chemical change has taken place.
Methane gas burns to form CO_2 and water.		Burning signifies a chemical change.

1.61 a) A molecule is composed of two or more nonmetallic atoms attached or bonded to one another in a specific way. Examples include H_2 and CO_2. A formula unit represents a unit of the formula written. If the element or elements in the formula are all nonmetals, this unit is called a molecule. If the elements in the formula contain a metal, the unit is usually not referred to as a molecule. It is, then, referred to just as a formula unit. Examples include NaCl, $CaCO_3$, and MnO_2. The difference between molecule and formula unit is that molecule specifies you have a covalent substance.

b) A compound is a pure substance that consists of two or more elements in definite composition and definite proportion. A mixture is not a pure substance. It has variable composition and variable proportions. Examples of a compound include water, H_2O; carbon dioxide, CO_2; and sulfuric acid, H_2SO_4. Examples of a mixture include tea, air, and soup. Notice that a mixture is not definite. If we take tea, it can be strong or weak, sweetened or unsweetened, hot or iced, flavored or not, etc.

1.63 a) The identity and count of the elements present in $MgNH_4AsO_4$ are one magnesium, one nitrogen, four hydrogen, one arsenic, and four oxygen.

b) The identity and count of the elements present in $Na_2B_4O_7 \cdot 10\ H_2O$ are two sodium, four boron, seventeen oxygen, and twenty hydrogen.

1.65 a) Because a neutral atom has equal number of protons as electrons, this atom has 15 electrons.

b) The atomic number of this atom is 20, and the atomic mass is 41 (20 + 21). The symbol is ${}^{41}_{20}$Ca.

c) The atomic number of this atom is 35, and the atomic mass is 80(35 + 45). The symbol is ${}^{80}_{35}$Br.

1.67 Gold is a metal. It is malleable, bends easily, and ductile—it can be drawn into wire.

1.69 A physical property is one that characterizes the physical existence, that is, state. A property that deals with reaction of a substance, that is, burning, is chemical. Burning (b) in air is a chemical, not physical, property.

1.71 a) Malleability is the property of bending or flattening into sheets.

b) Ductility is the property of drawing into wire.

1.73 a) Carbon sublines—becomes gas—at 3652°C. At 3675°C it will be a gas.

b) The boiling point of oxygen is −182.9°C. At 100°C it is a gas.

c) Carbon monoxide melts at −205°C. At −220°C it is a solid.

d) Methane melts at −182.6°C and boils at −161.4°C. At −175°C it will be a liquid.

1.75 High boiling point is a property of metals, whereas poor conductivity of heat and electricity is a property of nonmetals. A substance having both these properties is expected to be a semimetal.

1.77 Silicon is the element most likely to have a low boiling point and be a poor conductor because it is a nonmetal.

1.79 a) The alkali metal in period 6 is cesium, Cs.

b) The chalcogen in period 4 is selenium, Se.

c) The halogen in period 5 is iodine, I.

CHAPTER 2

2.1 The titles freshmen, sophomore, junior, senior, President *Pro Tempore,* and your Highness are all examples of "official" names from outside chemistry that seem strange.

2.3 You need to look for compounds containing only nonmetals or semimetals. The molecular compounds listed are:

SiH_4 SF_6 PCl_3 HI

2.5 Write the symbol for each element, then determine the subscript for each from the prefix.

a) nitrogen triiodide $\rightarrow NI_3$

b) diphosphorus pentaoxide $\rightarrow P_2O_5$

c) phosphorus tribomide $\rightarrow PBr_3$

d) dichlorine monoxide $\rightarrow Cl_2O$

e) sulfur tetrafluoride $\rightarrow SF_4$

2.7 All these compounds are made up of N and O—they are all nitrogen oxides. We distinguish between them by the prefixes used for each.

a) N_2O →dinitrogen monoxide

b) NO →nitrogen monoxide

c) N_2O_3 →dinitrogen trioxide

d) N_2O_4 →dinitrogen tetroxide

e) N_2O_5 →dinitrogen pentoxide

2.9 One of these compounds has a trivial name (see Chapter 2, page 50).

a) NH_3 →ammonia (nitrogen trihydride)

b) NF_3 →nitrogen trifluoride

2.11 See Web Animator.

2.13 Draw the dot structure for each atom, then put them together.

:Br· ·Br: → :Br–Br:

Two neutral Br atoms A Br_2 molecule

H· ·Br: → H–Br:

:Br· ·F: → :Br–F:

2.15 PH_3

P is the central atom with 3-H atoms around it.

$$\begin{array}{c} \mathrm{H}-\ddot{\mathrm{P}}-\mathrm{H} \\ | \\ \mathrm{H} \end{array}$$

NH_2^-

N is the central atom. There are eight electrons in this structure (5 + 2 + 1 = 8).

$$\left[\begin{array}{ccc} & \cdot\ddot{\mathrm{N}}\cdot & \\ / & & \backslash \\ \mathrm{H} & & \mathrm{H} \end{array}\right]^-$$

BH_4^-

B is the central atom with 4 H-atoms around it. There are eight electrons in this structure ((3 + 4) + 1 = 8).

$$\left[\begin{array}{c} \mathrm{H} \\ | \\ \mathrm{H}-\mathrm{B}-\mathrm{H} \\ | \\ \mathrm{H} \end{array}\right]^-$$

CH_3^-

C is the central atom. There are eight electrons in this structure (4 + 3 + 1 = 8).

$$\left[\begin{array}{c} \mathrm{H}-\ddot{\mathrm{C}}-\mathrm{H} \\ | \\ \mathrm{H} \end{array}\right]^-$$

2.17 a) See Example 2.7

$$:\mathrm{C}\equiv\mathrm{O}:$$

b) H_2S

2 + 6 = 8 electrons, S is central.

$$\mathrm{H}-\ddot{\underset{\cdot\cdot}{\mathrm{S}}}-\mathrm{H}$$

c) HCN

1 + 4 + 5 = 10 electrons, C is central.

$$\mathrm{H}\cdot\ \cdot\dot{\underset{\cdot}{\mathrm{C}}}\cdot\ \cdot\ddot{\mathrm{N}}:$$

Assume single bonding between the atoms keeping in mind that there is a total electron count of 10. Hydrogen already has its 2 electrons, nitrogen has 8, and carbon has 4. Convert the two lone pairs from nitrogen into bonds between nitrogen and carbon.

H−C≡N:

2.19 You need to draw one structure first.

6 + 3(6) = 24 e⁻, S is central.

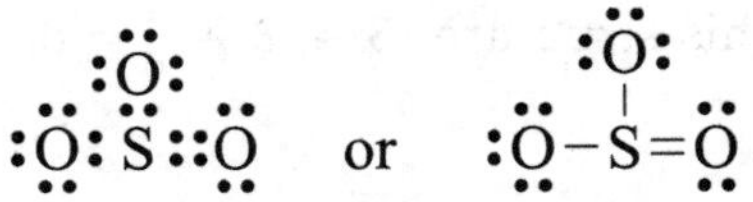

```
   ..              ..
  :O:             :O:
 ..  ..  ..   or   ..   |   ..
:O: S ::O        :O − S = O
 ..      ..        ..       ..
```

Now we can show the double bond in three different places.

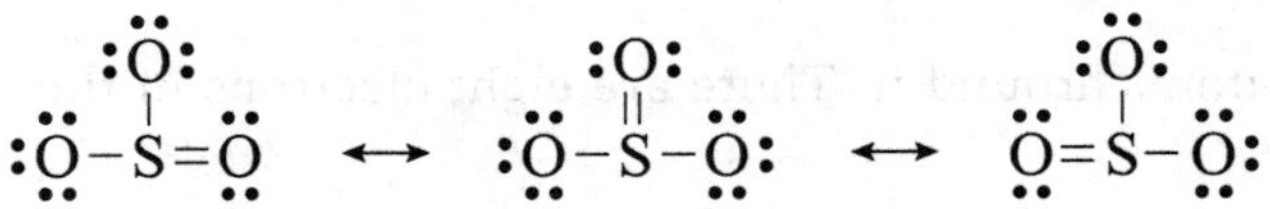

```
   ..                 :O:                  ..
  :O:                  ||                 :O:
 ..  |  ..       ..        ..         ..   |   ..
:O − S = O  ↔   :O − S − O:   ↔      O = S − O:
 ..      ..      ..        ..         ..       ..
```

2.21 SiO_4^{4-}

4 + 4(6) + 4 = 32e⁻

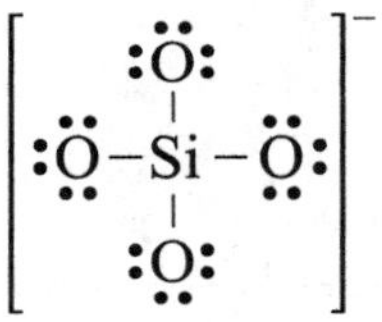

```
[      ..      ]−
[     :O:      ]
[ ..   |    .. ]
[:O − Si − O:  ]
[ ..   |    .. ]
[     :O:      ]
[      ..      ]
```

PO_4^{3-}

5 + 4(6) + 3 = 32e⁻

```
[      ..      ]3−
[     :O:      ]
[ ..   |   ..  ]
[:O − P − O:   ]
[ ..   |   ..  ]
[     :O:      ]
[      ..      ]
```

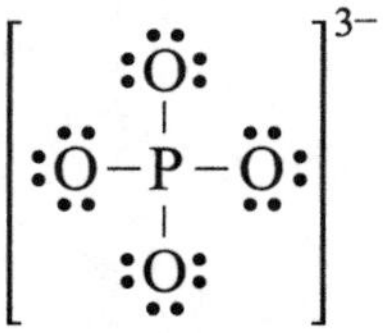

SO_4^{2-}

6 + 4(6) + 2 = 32e⁻

```
[      ..      ]2−
[     :O:      ]
[ ..   |   ..  ]
[:O − S − O:   ]
[ ..   |   ..  ]
[     :O:      ]
[      ..      ]
```

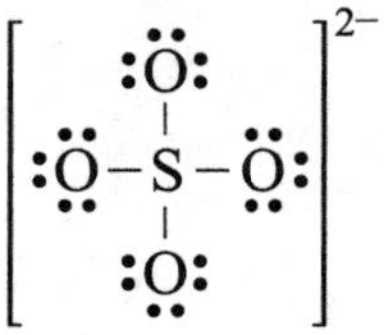

ClO_4^-

$7 + 4(6) + 1 = 32e^-$

```
      ••
     :O:
      |
 :O – Cl – O:
      |
     :O:
      ••          ]⁻
```

All ions have a central atom and 4 oxygens; giving us a count of 32 valence electrons; hence, they are all isoelectronic having the same number of electrons.

2.23 a) H_2Se has the same structure as H_2O and H_2S.

$2(1) + 6 = 8e^-$, Se is central.

```
     ••
H – :Se – H
```

b) SO_2 S is central with two oxygens bonded to it.

$6 + 2(6) = 18e^-$ (See problem 17f.)

```
 ••   ••   ••
:O – S = O
 ••        ••
```

2.25 a) ClO_3^-

$7 + 3(6) = 26e^-$, Cl is central

```
[ :O – Cl – O: ]⁻
        |
       :O:
```

Cl has to bond with three oxygens.

b) PF_4^+

$5 + 4(7) - 1 = 32e^-$

```
       ••
      :F:
       |
 :F – P – F:
       |
      :F:
       ••        ]⁺
```

This structure will be similar to those in problem 21.

c) NO_2^-

$5 + 2(6) + 1 = 18e^-$, N is central

N can form three bonds.

```
[ :O – N = O: ]⁻
```

2.27 BrO^-

$7 + 6 + 1 = 14e^-$

$[:\ddot{\underset{..}{Br}}-\ddot{\underset{..}{O}}:]^-$

The subsequent ions are formed when the O atom bonds with a lone pair of electrons from Br.

BrO_2^-

```
[ ..   ..  ]-
[ :Br – O: ]
[  |   ..  ]
[ :O:      ]
[  ..      ]
```

BrO_3^-

```
[ ..   ..   ..  ]-
[ :O – Br – O:  ]
[ ..   |    ..  ]
[     :O:       ]
[      ..       ]
```

BrO_4^-

```
[      ..       ]-
[     :O:       ]
[ ..   |    ..  ]
[ :O – Br – O:  ]
[ ..   |    ..  ]
[     :O:       ]
[      ..       ]
```

BrO_5^- should only have 38 valence electrons, and this figure has 40. There are no more lone pairs on Br, hence BrO_5^- will not form.

```
[       ..        ]-
[ ..   :O:   ..   ]
[ :O    |    O:   ]
[ .. \  |  / ..   ]
[      Br         ]
[     /    \      ]
[  :O:     :O:    ]
[   ..      ..    ]
```

2.29 See Web Animator on *The Practice of Chemistry* Web site.

2.31 a) N^+ $5 - 1 = 4e^-$

b) C^{2-} $4 + 2 = 6e^-$

c) N^- $5 + 1 = 6e^-$

d) Si^{4-} $4 + 4 = 8e^-$

e) H^- $1 + 1 = 2e^-$

f) Br^+ $7 - 1 = 6e^-$

g) B^{3-} $3 + 3 = 6e^-$

B^{3-} is not possible. It is unlikely that boron will gain three electrons seeing as boron is so small. The electron repulsion around the B would be too great.

2.33 AlF_4^- $3 + 4(7) + 1 = 32e^-$

This will have the same structure as the ions in problems 21 and 25b.

$$\left[\begin{array}{c} :\ddot{F}: \\ | \\ :\ddot{F}-Al-\ddot{F}: \\ | \\ :\ddot{F}: \end{array}\right]^-$$

2.35 (See page 38, Chapter 2 for table.)

Structure	Skeleton OK	e^- in structure	e^- from the atoms	e^- on outer atoms	e^- on central atoms
XeO_4 (Xe double-bonded to four O)	Yes	$32e^-$	Xe = 8 O = 4(6) Total = $32e^-$	$8e^-$ each	$16e^-$
$:\ddot{F}-N=\ddot{F}$ with $:\ddot{F}:$ bonded to N	No	$24e^-$	N = 5 F = 3(7) Total = $26e^-$	$8e^-$ each	$8e^-$
$\ddot{O}=\ddot{C}=\ddot{O}$	No	$18e^-$	C = 4 O = 2(6) Total = $16e^-$	$8e^-$ each	$10e^-$
$\ddot{S}=\ddot{O}-\ddot{O}:$	Yes	$18e^-$	S = 6 O = 2(6) Total = $18e^-$	$8e^-$ each	$8e^-$

2.37 SF_4

S has an expanded octet. This is possible because S is in the third period. There are single bonds to F.

SO_2

All atoms have an octet. This is a valid structure.

OF_2

F will form only one bond. This structure is invalid because there should be 20 electrons and there are only 18. The correct structure is:

$$:\ddot{F}-\ddot{O}-\ddot{F}:$$

2.39 a) OF_2—oxygen difluoride

b) N_2O_4—dinitrogen tetraoxide

c) CS_2—carbon disulfide

2.41 a) CBr_4

C is the central atom with 4-Br atoms around it. There are 4 + 4(7) = 32 electrons in this structure.

```
      ..                     ..
     :Br:                   :Br:
  .. .. ..               ..  |   ..
 :Br:C:Br:      or      :Br—C—Br:
  .. .. ..               ..  |   ..
     :Br:                   :Br:
      ..                     ..
```

b) $SiCl_4$

This structure is very similar to CBr_4 with 4 + 4(7) = 32 electrons in it.

```
      ..                     ..
     :Cl:                   :Cl:
  .. .. ..               ..  |   ..
 :Cl:Si:Cl:     or      :Cl—Si—Cl:
  .. .. ..               ..  |   ..
     :Cl:                   :Cl:
      ..                     ..
```

c) SO_2

S is the central atom in this structure. There are 6 + 2(6) = 18 electrons in this structure.

```
   ..  ..  ..            ..  ..  ..
  :O:S::O       or      :O—S=O
   ..      ..            ..      ..
```

2.43 a) antimony pentafluoride—SbF_5

b) diantimony trioxide—Sb_2O_3

2.45 Cl_2O_7 is dichlorine heptoxide. ClO_2 is chlorine dioxide. Cl_2O is dichlorine monoxide.

2.47 Selenium is in group VI, hence it has six valence electrons. Selenium has a total of 34 electrons, so there are 28 core electrons in an atom of selenium. Iodine has 53 electrons—7 valence and 46 core.

2.49 KF is an ionic compound and is expected to be a solid under normal conditions. NCl_3 and SO_3 are molecular or covalent compounds. They would be expected to form a gas under normal conditions.

2.51 The components of a Lewis structure are two or more atoms, represented by their symbols, and their valence electrons, represented by dots.

2.53 There are a total of 3 + 2 = 5 pairs, or 10 electrons in carbon monoxide.

2.55 There are 7 + 6 + 1 = 14 valence electrons in ClO^- and 7 + 4(6) + 1 = 32 valence electrons in ClO_4^-.

2.57 O_2^- has 2(6) + 1 = 13 valence electrons.

```
   ..   ..   -
 [:O — O:]
   .    ..
```

2.59 N_2O 2(5) + 6 = 16 electrons

```
 ..      ..
 O=N=N
 ..      ..
```

Another structure can be drawn with a triple bond.

```
  ..
 :O—N≡N:
  ..
```

CHAPTER 3

3.1 Ga is in group III and has 3 valence electrons, and As is in group V and therefore has 5 valence electrons. In (group III) has 3 valence electrons and P (group V) has 5 valence electrons. Cd is a transition metal with 2 valence electrons, and S (group 6) has 6 valence electrons.

One pattern lies in the fact that Si (group IV) has 4 valence electrons. When the above elements come together there is a total of $8e^-$ for the two atoms—an average of 4 valence electrons (just like Si) each. Also, the atoms in each pair are "equidistant" from Si group VI and they are all within two groups of silicon.

3.3 "The ions of nonmetals tend to have eight valence electrons."

3.5 To determine the number of valence electrons in an ion you must start with the number of valence electrons in an atom and either add (for anions) or subtract (for cations) the electrons according to the charge.

Ion	Valence electron in ion
Sn^{2+}	$4 - 2 = 2$
Mg^{2+}	$2 - 2 = 0$
Ca^{2+}	$2 - 2 = 0$
C^{3-}	$4 + 3 = 7$
Sn^{2-}	$6 + 2 = 8$
P^{3+}	$5 - 3 = 2$

3.7 See Table 3-3.

a) Li^+ Se^{2-} ∴ Li_2Se is correct.

b) Sr^{2+} Cl^- ∴ $SrCl_2$ not SrCl.

c) K^+ Br^- ∴ KBr not KBr_2.

d) Al^{3+} O^{2-} ∴ Al_2O_3 not Al_3O_3.

e) Ag^+ I^- ∴ AgI not AgI_2.

3.9 a) K_3N →potassium nitride

b) BaS →barium sulfide

c) CdO →cadmium oxide

3.11 See Table 3-3. When one polyatomic ion is needed, the ion is written as it is with no parentheses. When more than one polyatomic ion is needed to balance the charges, you must use parentheses around the polyatomic ion so you do not confuse the subscript number in the ion with the number of ions needed in the formulas.

a) Need two Na^{2+} for each CrO_4^{2} →Na_2CrO_4

b) Need two $C_2H_3O_2^-$ for each Ba^{2+} →must use parentheses for $C_2H_3O_2^-$

$$Ba(C_2H_3O_2)_2$$

c) Need three NO_2^- for each Al^{3+} →must use parentheses

$$Al(NO_2)_3$$

3.13 a) potassium permanganate
K^+ MnO_4^- $\rightarrow KMnO_4$
b) cadmium cyanide
Cd^{2+} CN^- $\rightarrow Cd(CN)_2$
c) silver nitrate
Ag^+ NO_3^- $\rightarrow AgNO_3$

3.15 a) sulfate ion $\rightarrow SO_4^{2-}$
b) sulfite ion $\rightarrow SO_3^{2-}$
c) nitrate ion $\rightarrow NO_3^-$
d) nitrite ion $\rightarrow NO_2^-$
e) phosphate ion $\rightarrow PO_4^{3-}$
f) phosphite ion $\rightarrow PO_3^{3-}$

3.17 See Table 3-5.
a) potassium oxalate
b) silver chromate
c) aluminum carbonate
d) magnesium perbromate

3.19 1) Using "black" and "white" for race is a conventional way of labeling people for census purposes, but it is not entirely correct.
2) Using directions of north, south, east or west for roads when the road may not truly be heading due north, south, east or west.
3) In chemistry, it is conventional to draw Lewis structures that imply that electrons sit in pairs around atoms to form bonds—but we know that electrons do not just hover around electrons forming bonds the way they are depicted according to Lewis Dot theory. Also, the notion that ionic substances contain actual full positive and negative charges is conventional, not factual.

3.21 See Table 3-6.
a) CsO_2 Cs = +1; O = −1/2
b) NaS_5 Na = +1, S = −1/5
c) Pb_3O_4 Pb = +8/3; O = −2

3.23 To determine the oxidation number of the metal you have to look at the oxidation number of the nonmetal and make sure there is charge balance.
a) Pb^{2+} (I_2^-)
b) Sn^{4+} (O_2^{2-})
c) Cu_2^+ (O^{2-})
d) Fe_2^{3+} (S_3^{2-})

3.25

Compound	Charge on ions	Number of valence electrons in ions
PbO_2	Pb^{4+} O^{2-}	Pb^{4+} : 0; O^{2-} : 8
PbO	Pb^{2+} O^{2-}	Pb^{2+} : 2; O^{2-} : 8
In_2Se_3	In^{3+} Se^{2-}	In^{3+} : 0; Se^{2-}: 8
InI_3	In^{3+} I^-	In^{3+} : 0; I^- : 8
$GaCl_3$	Ga^{3+} Cl^-	Ga^{3+} : 0; Cl^- : 8

3.27 a) iron (II) nitrate
b) aluminum perchlorate
c) calcium hypochlorite
d) uranium (IV) sulfate

3.29 Start with the charges.
a) Silver nitrate is $AgNO_3$.
b) Plutonium (IV) sulfate is $Pu(SO_4)_2$.
c) Lead (II) sulfate is $PbSO_4$.

3.31 a) $K^{2+}Cr_2{}^{6+}O_7{}^{2-}$
b) $Sn^{2+}Br_2{}^{-}$
c) We know that nitrate ($NO_3{}^-$) has a charge of −1. This means the cobalt has an oxidation number of +3. You can figure out the charge of N algebraically.
$Co^{3+}(N^{5+}O_3{}^{2-})_3$
d) $Mn^{2+}Cr^{6+}O_4{}^{2-}$

3.33

Compound	Cation	Anion	Name of compound
$Fe(NO_3)_3$	Fe^{3+}, iron (III)	$NO_3{}^-$, nitrate	iron (III) nitrate
$CuSO_4$	Cu^{2+}, copper (II)	$SO_4{}^{2-}$, sulfate	copper (II) sulfate
Li_2CO_3	Li^+, lithium	$CO_3{}^{2-}$, carbonate	lithium carbonate
K_3N	K^+, potassium	N^{3-}, nitride	potassium nitride
$Sn(OH)_2$	Sn^{2+}, tin (II)	OH^-, hydroxide	tin (II) hydroxide

3.35

	Cl^-	$S_2{}^-$	$N_3{}^-$	F^-	$NO_3{}^-$	$SO_4{}^{2-}$	$PO_4{}^{3-}$
Na^+	$NaCl$	Na_2S	Na_3N	NaF	$NaNO_3$	Na_2SO_4	Na_3PO_4
Mg^{2+}	$MgCl_2$	MgS	Mg_3N_2	MgF_2	$Mg(NO_3)_2$	$MgSO_4$	$Mg_3(PO_4)_2$
Al^{3+}	$AlCl_3$	Al_2S_3	AlN	AlF_3	$Al(NO_3)_3$	$Al_2(SO_4)_3$	$AlPO_4$
NH^{4+}	NH_4Cl	$(NH_4)_2S$	$(NH_4)_3N$	NH_4F_3	NH_4NO_3	$(NH_4)_2SO_4$	$(NH_4)_3PO_4$
Cr^{3+}	$CrCl_3$	Cr_2S_3	CrN	CrF_3	$Cr(NO_3)_3$	$Cr_2(SO_4)_3$	$CrPO_4$
Cu^{2+}	$CuCl_2$	CuS	Cu_3N_2	CuF_2	$Cu(NO_3)_2$	$CuSO_4$	$Cu_3(PO_4)_2$

3.37 a) barium sulfite
b) calcium chlorite
c) cadmium nitrite
d) aluminum acetate

3.39 a) Cr^{3+}; $Cl_3{}^-$
b) Mn^{4+}; $O_2{}^{2-}$
c) $Cu_2{}^+$; O^{2-}
d) Cu^{2+}; O^{2-}
e) Zn^{2+}; $F_2{}^-$

3.41 a) Mn^{4+} (O_2^{2-}); Manganese (IV) oxide
b) Co^{3+} (Cl_3^{-}); Cobalt (III) chloride
c) Ni^{+} (O^{2-}); Nickel (I) oxide
d) Cr^{2+} (F_2^{-}); Chromium (II) fluoride

3.43 See Problem 23. Remember, when polyatomic ions appear only once in a formula, we do not put parentheses around it. If more than one polyatomic ion is needed to balance the charges, put it in the parentheses and use a subscript outside the parentheses. This is so you do not confuse the final subscript in the polyatomic ion with the number of ions needed for charge balance.

a) $Na_3^{3+}PO_4^{3-}$
Sodium has an oxidation number of +1.
The compound is sodium phosphate.

b) $Ti_2^{3+}(CO_3)_3^{2-}$
Titanium has an oxidation number of +3.
The compound is titanium (III) carbonate.

c) $Mn^{2+}SO_4^{2-}$
Manganese has an oxidation number of +2.
The compound is manganese (II) sulfate.

d) $K^{+}NO_2^{-}$
Potassium has an oxidation number of +1.
The compound is potassium nitrite.

3.45 Start with the charges.
a) Calcium sulfite is $CaSO_3$.
b) Iron (II) carbonate is $FeCO_3$.
c) Aluminum hydrogen phosphate is $Al_2(HPO_4)_3$.

3.47 S and O are both chalcogens. When they combine with another element, they will both exhibit the same charge. This is supported by the periodic law. Because Ga is in the same group as Al, it will form analogous compounds with O and S—Ga_2O_3 and Ga_2S_3.

3.49 P and As are in the same group. The polyatomic ions formed by combining As and O are AsO_4^{3-} and AsO_3^{3-}.

3.51 a) magnesium hypoiodite—$Mg(IO)_2$
b) manganese (III) dichromate—$Mn_2(Cr_2O_7)_3$
c) magnesium permanganate—$Mg(MnO_4)_2$

3.53 a) Co^{2+} $(N^{5+}$ $O^{2-}{}_3)_2$
b) Pt^{4+} $(S^{6+}$ $O^{2-}{}_4)_2$

3.55 See problem 3.20 and Table 3-6.
a) Ag_2^{+} Cr_2^{6+} O_7^{2-}
b) Ba^{2+} Cr_2^{6+} O_7^{2-}
c) Al_2^{3+} $(Cr_2^{6+}$ $O_7^{2-})_3$

3.57 a) cesium superoxide—CsO_2
b) potassium superoxide—KO_2

CHAPTER 4

4.1 1 mole = 6.02×10^{23} molecules

0.5 mole = 3.01×10^{23} molecules

It is not possible to have a fraction of a molecule. A fraction of a mole, however, is a whole number of molecules. Therefore, it is acceptable to have "half a mole."

4.3 a) For every dozen cakes, you will need three dozen eggs.

b) When you make sandwiches for six dozen students, you will need one-half dozen loaves of bread (or six loaves).

c) Building a dozen houses requires about one thousand dozens (12,000 pieces) of wood.

4.5 a) We find that there are three iron atoms in one formula unit of iron (III) oxide.

b) For the molecular substance sulfur hexafluoride there are six fluorine atoms for every sulfur atom.

c) "White" phosphorus is a molecular substance with four atoms of phosphorus per molecule of white phosphorus.

4.7 a) $\frac{\text{3 Br atoms}}{\text{1 molecule}}$

b) $\frac{\text{1 P atom}}{\text{1 molecule}}$

c) $\frac{\text{13 H atoms}}{\text{1 molecule}}$

d) $\frac{\text{4 Si atoms}}{\text{1 molecule}}$

4.9 a) $\frac{\text{1 mole Pt + 6 moles F}}{\text{1 mole compound}}$ gives the formula PtF_6

PtF_6 gives rise to the ratios 1 mole Pt/1 mole PtF_6 and 6 mole F/1 mole PtF_6.

b) $\frac{\text{3 moles Ti + 4 moles } PO_4}{\text{1 mole compound}} \rightarrow Ti_3(PO_4)_4$

$Ti_3(PO_4)_4$ gives rise to the ratios 3 moles Ti/ 1 mole $Ti_3(PO_4)_4$, 4 moles P/ 1 mole $Ti_3(PO_4)_4$. And 16 moles O/ 1 mole $Ti_3(PO_4)_4$.

c) $\frac{\text{1 mole Fe + 1 mole } NH_4 \text{ + 2 mole } SO_4}{\text{1 mole compound}} \rightarrow FeNH_4(SO_4)_2$

$FeNH_4(SO_4)_2$ gives rise to the ratios 1 mole Fe/ 1 mole $FeNH_4(SO_4)_2$, 1 mole N/ 1 mole $FeNH_4(SO_4)_2$, 4 moles H/ 1 mole $FeNH_4(SO_4)_2$, 2 moles S/ 1 mole $FeNH_4(SO_4)_2$, and 8 moles O/ 1 mole $FeNH_4(SO_4)_2$.

4.11 $\frac{1200 \text{ atoms H}}{1 \text{ molecule protein}}$ *then* $\frac{1200 \text{ moles H atoms}}{1 \text{ mole protein molecule}}$

4.13 "Magnesium metal reacts with oxygen gas" ... This means that magnesium and oxygen are reactants. And "... to form magnesium oxide" means that magnesium oxide is the product.

Reactants = Mg, O_2

Product = MgO

4.15 The reactants are aluminum metal and fluorine, and the product is aluminum fluoride.

Reactants = Al, F_2

Product = AlF_3

4.17 $CH_4 + 2\ O_2 \rightarrow CO_2 + 2\ H_2O$

To correctly represent the numbers indicated in the equation, we need to draw two O_2 and two H_2O.

```
   H        :Ö=Ö:                        Ö
   |                                    / \
H–C–H  +               →  Ö=C=Ö  +     H   H
   |
   H        :Ö=Ö:                        Ö
                                        / \
                                       H   H
```

We can see there is one carbon, four hydrogen, and four oxygen on both sides.

4.19 $H_2 + Cl_2 \rightarrow 2\ HCl$

```
H–H  +  :C̈l–C̈l:  →  H–C̈l:  +  H–C̈l:
```

We can see that there are two hydrogen and two chlorine on both sides.

4.21 a) To make one formula unit of copper (II) nitrate from copper requires two molecules of nitric acid, HNO_3.

b) When we get two molecules of oxygen by the decomposition of hydrogen peroxide, we also form a molecule of water.

c) Conversion of two molecules of methane to one molecule of ethanol also requires one molecule of oxygen.

4.23 a) See 4.22(a). The method of balancing is the same; 2 O's on the right and only 1 on the left. This puts a 2 in front of the HgO and finally a 2 in front of the Hg.

$$2HgO \rightarrow 2Hg + O_2$$

b) The H_2O makes an odd number of oxygen in the products. We need to double the H_2O to make an even number of oxygen in the products then balance the equation.

$$2H_2O_2 \rightarrow 2H_2O + O_2$$

4.25 a) We can balance NO_3^- as a unit without breaking it down to nitrogen and oxygen because it stays intact in the product.

$$Cu + 2AgNO_3 \rightarrow 2Ag + Cu(NO_3)_2$$

b) For the NO_3^- to balance, the least common multiple of 2 and 3 is 6. We need to get 6 molecules of NO_3^- on both sides.

$$2Al + 3Sn(NO_3)_2 \rightarrow 2Al(NO_3)_3 + 3Sn$$

c) NH_4^+, $C_2H_3O_2^-$, and PO_4^{3-} all can be treated as units since they stay intact in the product side. If we balance barium first the rest will fall into place.

$$3Ba(C_2H_3O_2)_2 + 2(NH_4)_3PO_4 \rightarrow Ba_3(PO_4)_2 + 6NH_4C_2H_3O_2$$

4.27 a) We find that there are two carbon atoms in one ethane molecule.

$$\frac{\text{2 atoms carbon}}{\text{1 molecule ethane}}$$

b) Decomposition of germane, GeH_4, yields two molecules of hydrogen gas for every germane molecule.

$$\frac{\text{2 molecules hydrogen}}{\text{1 molecule germane}}$$

c) The destruction of nitrogen dioxide in a car's catalytic converter yields one molecule of nitrogen gas for two molecules of nitrogen dioxide.

$$\frac{\text{1 molecule nitrogen}}{\text{2 molecules nitrogen dioxide}}$$

4.29

$$\frac{\text{4 atoms Fe}}{\text{1 unit hemoglobin}} \rightarrow \frac{\text{4 moles Fe}}{\text{1 mole hemoglobin}}$$

4.31 a) In order to balance this equation, we have to double the amount of iron and triple the amount of sulfate. Now we can get a total count of H in the reactants (12) and balance the equation by putting a 6 in front of the H_2O.

$$2Fe(OH)_3 + 3H_2SO_4 \rightarrow Fe_2(SO_4)_3 + 6H_2O$$

b) In order to balance this equation, we first balance the Cl by putting a 2 in front of NaCl. This requires us to also put a 2 in front of NaOH to balance the Na, and finally, we put a 2 in front of the H_2O to balance the H.

$$2NaCl + 2H_2O \rightarrow Cl_2 + H_2 + 2NaOH$$

c) In order to balance this equation, we first balance the Na by putting a 2 in front of NaCl. This requires that we also put a 2 in front of the H_2O to balance the oxygen. The equation is now balanced.

$$Na_2NH + 2H_2O \rightarrow NH_3 + 2NaOH$$

d) In this equation, both Al and O appear in more than one formula on the right; we will balance these last. We begin by balancing the Cl by putting a 3 in front of NH_4ClO_4. This requires us to also put a 3 in front of NO to balance N; we also put a 6 in front of H_2O to balance H. With one Al in $AlCl_3$ and two in Al_2O_3, we need to put a 3 in front of the reactant Al. We see that O is now balanced (a total of 12 on the left and 12 on the right).

$$3Al + 3NH_4ClO_4 \rightarrow Al_2O_3 + AlCl_3 + 3NO + 6H_2O$$

e) In order to balance this equation, we first balance F by putting a 4 in front of NaF. This requires us to also put a 4 in front of NaCl to balance Na. We now have a total of 6 Cl and 3 S on the right; putting a 3 in front of SCl_2 balances the equation.

$$3SCl_2 + 4NaF \rightarrow SF_4 + S_2Cl_2 + 4NaCl$$

f) In order to balance this equation, we first balance F by putting a 4 in front of HF. This requires us to put a 2 in front of H_2O to balance H.

$$UO_2 + 4HF \rightarrow UF_4 + 2H_2O$$

g) In order to balance this equation, we first balance H by putting a 2 in front of HNO_3. The equation is now balanced.

$$N_2O_5 + H_2O \rightarrow 2HNO_3$$

4.33 a)

$$4\,H{-}\underset{H}{\overset{..}{N}}{-}H + 5\,O{=}O \rightarrow 4\,N{=}O + 6\,H{-}O{-}H$$

There are four N, ten O, and twelve H.

b)

$$2H{-}\underset{H}{\overset{H}{C}}{-}\underset{H}{\overset{H}{C}}{-}\underset{H}{\overset{H}{C}}{-}\underset{H}{\overset{H}{C}}{-}\underset{H}{\overset{H}{C}}{-}H + 15\,O{=}O \rightarrow 10\,O{=}C{=}O + 10\,H{-}O{-}H$$

There are ten C, twenty H, and thirty O.

c)

$$2\,O{=}S{=}O + O{=}O \rightarrow 2\,SO_3$$

There are two S and six O.

4.35 a) potassium bromide + chlorine gas → potassium chloride + bromine

b) hydrogen chloride + ammonia gas → ammonium chloride

c) magnesium sulfide + sulfuric acid → magnesium sulfate + hydrogen sulfide

d) potassium carbonate + hydrogen chloride → carbon dioxide + water + potassium chloride

4.37 a) $2KBr(s) + Cl_2(g) \rightarrow 2KCl(s) + Br_2(l)$

b) $HCl(g) + NH_3(g) \rightarrow NH_4Cl(s)$

c) $MgS(s) + H_2SO_4(aq) \rightarrow MgSO_4(s) + H_2S(g)$

d) $K_2CO_3(s) + 2HCl(l) \rightarrow CO_2(g) + H_2O(l) + 2KCl(s)$

4.39 a) C_5H_{12}

b) $AgNO_3$

c) NO_2O_4 We have to adjust the ratio 1 mole N/0.5 mole compound; 2 mole 0/0.5 mole compound so that we have whole numbers. To do this, multiply the numerator and denominator by 2.

4.41 $Mn_2^{7+}\ O_4^{2-}$ Manganese (VII) oxide

4.43 Reducing this ratio gives us MnO_2, manganese (IV) oxide.

4.45 $2\,C + O{=}O \rightarrow 2\,C{\equiv}O$

4.47 $6P + 5KClO_3 \rightarrow 5KCl + 3P_2O_5$

4.49 a) We can get whole number coefficients by multiplying through by 2.

$$4Al + 3O_2 \rightarrow 2Al_2O_3$$

b) To get the simplest ratio we must divide through by 3.

$$Mg(NO_3)_2 + 2NaCl \rightarrow 2NaNO_3 + MgCl_2$$

c) We must multiply through by 3.

$$2Al + 6HCl \rightarrow 2AlCl_3 + 3H_2$$

4.51 a) 1 formula unit = $KClO_3$

There are 1 K-atom, 1 Cl-atom, and 3 O-atoms.

b) 1 formula unit = $Al_2(SO_4)_3$

There are 2 Al-atoms, 3 S-atoms, and 12 O-atoms.

c) 1 formula unit = $U_3(PO_3)_4$

There are 3 U-atoms, 4 P-atoms, and 12 O-atoms.

d) 1 formula unit = $(NH_4)_2CO_3$

There are 2 N-atoms, 8 H-atoms, 1 C-atom, and 3 O-atoms.

4.53 The balanced equation is

$16KClO_3 + 3S_8 \rightarrow 24SO_2 + 16KCl$

There is a ratio of 3 atoms S_8 to 2 formula units $KClO_3$, or $\frac{3}{2}$ atom S_8 to 1 formula unit $KClO_3$.

4.55 $3P_4S_3 + 16KClO_3 \rightarrow 6P_2O_5 + 9SO_2 + 16KCl$

a) P_4S_3—tetraphosphorus trisulfide

$KClO_3$—potassium chlorate

P_2O_5—diphosphorus pentoxide

SO_2—sulfur dioxide

KCl—potassium chloride

b) We assign the oxidation number of the most negative element first then figure out the other one. S has an oxidation number of −2 in P_4S_3 and +4 in SO_2.

$P_4^{3/2+}\,S_3^{2-}$; $S^{4+}\,O_2^{2-}$

c) K^+Cl^- ; $K^+Cl^{5+}O_3^{2-}$

Cl has an oxidation number of −1 in KCl and +5 in $KClO_3$.

CHAPTER 5

5.1
- A physical property is one that describes the physical existence of a substance—one that can be measured without changing the identity of that substance.
- A chemical property is one that describes how that substance reacts or changes its identity, thereby forming a new substance.

5.3 Both sodium and chlorine will readily react with oxygen because they are very reactive. Neither are ever found free (as elements) in nature. Also, sodium must be kept away from air and water (both sources of oxygen), which implies that it will react with oxygen, and chlorine is known to readily combine with all elements.

5.5 Physical properties describe the physical existence of an element or the changes it undergoes. Melting point, boiling point, density, physical state at 25°C, and physical appearance are all physical properties of an element.

5.7 a) First, write the chemical formulas for each substance:

acetic acid = $HC_2H_3O_2$

calcium carbonate = $CaCO_3$

calcium acetate = $Ca(C_2H_3O_2)_2$

water = H_2O

carbon dioxide = CO_2

Next, determine what the products and reactants will be:

Reactants = $HC_2H_3O_2$, $CaCO_3$

Products = $Ca(C_2H_3O_2)_2$, H_2O, and CO_2

$HC_2H_3O_2 + CaCO_3 \rightarrow Ca(C_2H_3O_2)_2 + H_2O + CO_2$

$C_2H_3O_2$ is not balanced; we need to balance it by doubling the amount of the reactant. Hydrogen is also not balanced.

$2C_2H_3O_2(aq) + CaCO_3(s) \rightarrow Ca(C_2H_3O_2)_2(aq) + H_2O(l) + CO_2(g)$.

b) There is a solid before the reaction ($CaCO_3$). When reaction occurs, a gas is formed (CO_2) and the solid disappears. The products are all soluble except for CO_2.

c) In order for CO_2 to form from $CaCO_3$ there must be acid present.

For problems 5.9 and 5.11 the only product possible in each is for the two elements to combine, forming a compound. Watch the charge balance between the two elements when they combine.

5.9 $2Cd + O_2 \rightarrow 2CdO$ Cd always has a charge of +2; O is −2.

5.11 a) $6Zn + P_4 \rightarrow 2Zn_3P_2$

b) $Ba + F_2 \rightarrow BaF_2$

c) $3Sr + N_2 \rightarrow Sr_3N_2$

d) $2Al + 3F_2 \rightarrow 2AlF_3$

5.13 (a) & (c) are synthesis reactions because in each two reactants come together to produce one substance. (b) is therefore decomposition since one reactant produces two substances in the products.

a) $4Bi + 3O_2 \rightarrow 2Bi_2O_3$

b) $Hg(OH)_2 \rightarrow HgO + H_2O$

c) $N_2O_3 + H_2O \rightarrow 2HNO_3$

5.15 a) See Example 5.8. Remember that Ag will have +1 charge when it combines.

$CuCl_2 + 2Ag \rightarrow 2AgCl + Cu$

b) See Example 5.6. Remember that bromine is diatomic.

$2KBr + Cl_2 \rightarrow 2KCl + Br_2$

c) $NiO + H_2 \rightarrow H_2O + Ni$

d) See Example 5.9.

$2Cr_2O_3 + 3C \rightarrow 4Cr + 3CO_2$

5.17 See Table 5-2.

a) lead (II) iodide + fluorine → single displacement

$PbI_2 + F_2 \rightarrow PbF_2 + I_2$

b) zinc oxide + hydrogen → single displacement

$ZnO + H_2 \rightarrow Zn + H_2O$

c) potassium + sulfur → synthesis

$16K + S_8 \rightarrow 8K_2S$

If we consider that the molecular formula of sulfur is S_8, then the balanced equation is shown above.

5.19 a) We should assign sulfur's oxidation state first then determine the oxidation state for Fe.

$Fe_2^{3+}S_3^{2-}$

b) The name of this compound is iron (III) sulfide, or ferric sulfide.

c) $16Fe + 3S_8 \rightarrow 8Fe_2S_3$

d) oxidized

$Fe^0 + S_8^0 \rightarrow Fe_2^{3+}S_3^{2-}$

reduced

e) $Fe_2S_3 + 2Al \rightarrow Al_2S_3 + 2Fe$

5.21 a) oxidized

$8Fe^0 + S_8^0 \rightarrow Fe^{2+}S^{2-}$

reduced

b) oxidized

$2Al^0 + 3F_2^0 \rightarrow 2Al^{3+}F_3^-$

reduced

5.23 See Table 5-3. If there is an insoluble compound produced in the reaction, the reaction will take place. If all products are soluble, there is no reaction.

a) $Cu_3(PO_4)_2$ is insoluble, so this reaction will take place.

b) $PbSO_4$ is insoluble, so this reaction will take place. Because there are solids on both sides, the reaction may only be partial.

c) $CaSO_4$ is insoluble; the reaction will take place. Because there are solids on both sides, the reaction may only be partial.

d) $BaSO_4$ is insoluble; the reaction will take place.

5.25 a) $Mg(NO_3)_2(aq) + 2NaOH(aq) \rightarrow 2NaNO_3(aq) + Mg(OH)_2(s)$

In this reaction, $Mg(OH)_2$ is insoluble.

b) $2Fe(C_2H_3O_2)_3(aq) + 3K_2SO_4(aq) \rightarrow Fe_2(SO_4)_3(s) + 6KC_2H_3O_2(aq)$

All substances are soluble.

c) $2Na_3PO_4(aq) + 3Cd(NO_3)_2(aq) \rightarrow Cd_3(PO_4)_2(s) + 6NaNO_3(aq)$

$Cd_3(PO_4)_2$ is insoluble.

5.27 $3ZnCl_2 + 2K_3PO_4 \rightarrow Zn_3(PO_4)_2 + 6KCl$

The insoluble substance is zinc phosphate.

5.29 a)

oxidized

$$C^{4-}H^{+}_{4} + Cl^{0}_{2} \rightarrow H^{+}Cl^{-} + C^{4+}Cl^{-}_{4}$$

reduced

b)

oxidized

$$O^{2+}F^{-}_{2} + H^{+}_{2}O^{2-} \rightarrow O^{0}_{2} + H^{+}F^{-}$$

reduced

c)

oxidized

$$N^{2-}_{2}H^{+}_{4} + 2H^{+}_{2}O^{-}_{2} \rightarrow N^{0}_{2} + 4H^{+}_{2}O^{2-}$$

reduced

5.31 a) $4C_3H_5O_9N_3 \rightarrow 12CO_2 + 6N_2 + O_2 + 10H_2O$

b) $(NH_4)_2Cr_2O_7 \rightarrow Cr_2O_3 + N_2 + 4H_2O$

5.33 a) $2AgNO_3 + Na_2CO_3 \rightarrow 2NaNO_3 + Ag_2CO_3$

b) $6HCl + Al_2O_3 \rightarrow 2AlCl_3 + 3H_2O$

c) $CaF_2 + H_2SO_4 \rightarrow 2HF + CaSO_4$

d) $2KI + Pb(NO_3)_2 \rightarrow 2KNO_3 + PbI_2$

5.35 Oxygen is in its free state and has oxidation number of 0. It can change its oxidation state, as in the case of combustion. Water has both hydrogen and oxygen in thin "reacted" oxidation states of +1 and −2, respectively. They will not undergo further change, putting out a fire rather than reacting further and "fueling" the fire.

5.37 The primary observation is that the product gas appears to be less flammable than the reactant gas. The product is a compound, most likely a non-metal oxide. One of the product's chemical properties is that it can be decomposed into oxygen and the original substance. Physical properties would include that it is a gas at room temperature and pressure; its density is less than 1.0 g/mL.

5.39 Since these are two elements, this will be a synthesis reaction. The product is most likely Sr_3N_2.

5.41 $H_2 + VO \rightarrow H_2VO_3$

5.43 $12K + P_4 \rightarrow 4K_3P$

5.45 Chemical reaction can result in formation of a gas (bubbles), formation of an insoluble substance (precipitate), change in color.

5.47 Combustion is the burning of a hydrocarbon, an organic carbon-based compound. It can be recognized as a hydrocarbon reacting with oxygen to produce carbon dioxide and water, usually.

5.49 a) Solid carbon dioxide sublimes to carbon dioxide gas.

b) $Na_2CO_3(s) + 2HCl(g) \rightarrow CO_2(g) + H_2O(l) + 2NaCl(aq)$

Solid sodium carbonate reacts with gaseous hydrogen chloride to produce a carbon dioxide gas, water, and aqueous sodium chloride.

5.51 a) $2AgCl + F_2 \rightarrow 2AgF + Cl_2$

b) $2GaBr_3 + 3Cl_2 \rightarrow 2GaCl_3 + 3Br_2$

c) $2Al_2O_3 + 3C \rightarrow 4Al + 3CO_2$

5.53 a) Pt has an oxidation state of +4 in $PtCl_4$.

b) $PtCl_4$ is platinum (IV) chloride.

c) $Pt + 2Cl_2 \rightarrow PtCl_4$

d) Pt is oxidized; Cl_2 is reduced

e) $PtCl_4 + 2F_2 \rightarrow PtF_4 + 2Cl_2$

CHAPTER 6

6.1 If you find the relationship between the broomstick, ballpoint pen, and paper clip, it will be easy to finish this table.

1 broomstick = 9 ballpoint pens = 36 paper clips

	Broomstick	Ballpoint pen	Paper clip
Shoe box	l = 1/6 w = 1/12 h = 1/12	l = 1 1/2 w = 3/4 h = 3/4	l = 6 w = 3 h = 3
Your kitchen	l = 3 w = 2 1/3	l = 27 w = 21	l = 108 w = 84

6.3 Intensive properties do not depend on size or amount. Extensive properties depend on size or amount.

a) Extensive property. The need for new school classrooms depends on the amount of housing.

b) Intensive property. The amount of fertilizer per acre depends on the crop.

c) Extensive property. The number of accidents depends on the number of vehicles on the road.

d) Intensive property. The lift does not depend on the size of the balloon.

6.5 There are 2.0 grams not accounted for after heating. It could be 2.0 grams of gas produced, therefore, 21.2 g material + 2.0 g gas = 23.2 g solid compound. This can be verified if the solid is heated in a test tube that has a stopper on it with tubing that goes into a cylinder filled with water. The gas would then displace the water and could be collected for analysis.

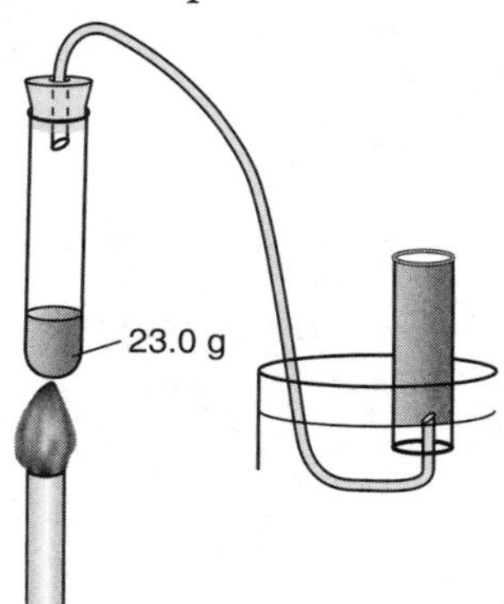

6.7 When the melted sulfur hits the cold water, it will cool down and freeze.

6.9 a) The temperatures, both boiling points and melting points, increase as you go down a group. Group III chlorides, $AlCl_3$, $GaCl_3$, and $InCl_3$, do not fit this trend.

b) Freezing points decrease as you go to the right. Boiling points seem to increase as you go right.

c) With room temperature at 30°C, substances with higher freezing points will be solid, ones with lower freezing points will be liquids, and ones with lower boiling points will be gas.

Solids: $AlCl_3$, $GaCl_3$, $InCl_3$

Liquids: $SiCl_4$, $GeCl_4$, $SnCl_4$, PCl_3, $AsCl_3$, $SbCl_3$

Gases: None

6.11 25.0 g

d = 1.59 g/cm³

Remember that density = mass/ volume (or d = m/v).

Therefore:

$$d = \frac{\text{mass}}{\text{volume}}$$

$$\text{volume} = \frac{\text{mass}}{d} = \frac{25.0\text{ g}}{1.59\text{ g/cm}^3} = 15.7\text{ cm}^3$$

6.13 $d_{air} = 0.00129\text{ g/cm}^3$

The balloon that contains gas with density less than 0.00129 g/cm³ will float.

a) Balloon 1 will not float.

b) Balloon 2 will float.

c) Balloon 3 will float.

d) Balloon 4 will float.

6.15

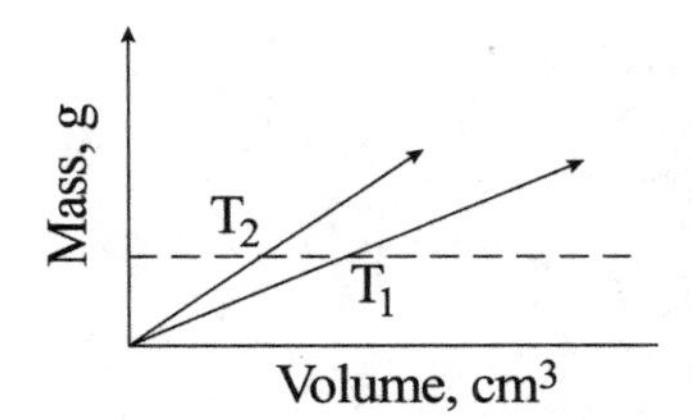

The substance's volume decreases at T_2. This means it contracts from T_1 to T_2.

6.17 When the temperature is raised from 25°C to 100°C, the substance will expand. This will cause the line for 100°C to be below the 25°C line.

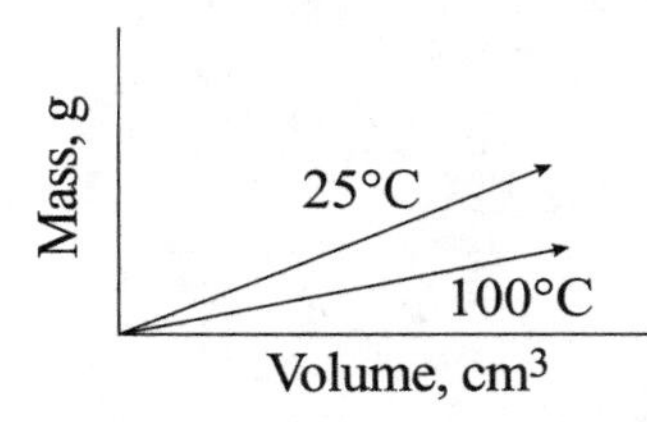

6.19 $$\text{volume} = \frac{\text{mass}}{d} = \frac{55\text{ g}}{0.985\,\frac{\text{g}}{\text{cm}^3}} = 55.8\text{ cm}^3 \approx 56\text{ cm}^3$$

6.21 $d = \frac{\text{mass}}{\text{volume}} = \frac{21.6\ g}{10.0\ cm^3} = 2.16\ \frac{g}{cm^3}$

6.23 Volume = 73.0 cm^3 – 40.0 cm^3 = 33.0 cm^3

$$d = \frac{60.0\ g}{33.0\ cm^3} = 1.82\ \frac{g}{cm^3}$$

6.25 a) miles, gallon
b) tons chlorine, tons NaOH
c) atoms carbon, methane molecule
d) molecules ammonia, molecule hydrogen

6.27 $\frac{300\ lb}{9\ in} = \frac{?}{ft}$

Since the proportion is in inches, you need to use 12 inches instead of 1 foot. Therefore, re-write the proportion:

$$\frac{300\ lb}{9\ in} = \frac{?}{12\ ft}$$

$$\frac{(12\ in)(300\ lb)}{(9\ in)} = 400\ lb$$

6.29 Percent means "per hundred."
65 = 12.2%

$$\frac{65}{?} = \frac{12.2\ \text{technicians}}{100\ \text{workers}}$$

$$\frac{(65)(100)}{12.2} = 533\ \text{workers}$$

6.31 3% copper, 23% silver, 28% mercury, 46% cobalt

$$0.253\ \cancel{g\ alloy} \times \frac{3\ g\ Cu}{100\ \cancel{g\ alloy}} = 0.8\ g\ Cu$$

$$0.253\ \cancel{g\ alloy} \times \frac{23\ g\ Ag}{100\ \cancel{g\ alloy}} = 0.058\ g\ Ag$$

$$0.253\ \cancel{g\ alloy} \times \frac{46\ g\ Co}{100\ \cancel{g\ alloy}} = 0.12\ g\ Co$$

$$0.253\ \cancel{g\ alloy} \times \frac{28\ g\ Hg}{100\ \cancel{g\ alloy}} = 0.071\ g\ Hg$$

6.33 $500.0 \text{ g ore} \times \frac{45 \text{ g Fe}}{100 \text{ g ore}} = 225.0 \text{ g Fe}$

6.35 The key to working through this problem is to see what is given and to write it all out. The problem gives you a 4.82% Ca, 500.0 mL solution, as well as a density of 1.25 g/mL. It asks you for the grams of calcium in the solution. If you write out what everything means, you have:

500 mL solution

$\frac{1.25 \text{ g solution}}{1 \text{ mL solution}}$ $\frac{4.82 \text{ g Ca}}{100 \text{ g solution}}$

Multiplying the given information together, we get the answer.

$$500.0 \text{ mL sol} \times \frac{1.25 \text{ g sol}}{1 \text{ mL sol}} \times \frac{4.82 \text{ g Ca}}{100 \text{ g sol}} = 30.1 \text{ g Ca}$$

6.37 $\frac{15.0 \text{ g } H_2}{71.0 \text{ g } O_2}$ so for 6.0 g H_2:

$$6.0 \text{ g } H^2 \times \frac{71.0 \text{ g } O_2}{15.0 \text{ g } H_2} = 28.4 \text{ g } O_2$$

6.39 $25 \text{ L bleach} \times \frac{100 \text{ L sol}}{10.0 \text{ L bleach}} = 250 \text{ L solution}$

6.41 $\frac{13 \text{ cm long}}{7 \text{ cm wide}} = \frac{? \text{ cm long}}{28 \text{ cm wide}}$

Answer: 52 cm long

6.43 $\frac{100{,}000 \text{ population}}{300 \text{ tagged}} = \frac{? \text{ population}}{1000 \text{ tagged}}$

Answer: Three alligators were tagged in the resample.

6.45 $275 \text{ formula units} \times \frac{12 \text{ atoms O}}{1 \text{ formula unit } Al_2(CrO_4)_3} = 3300 \text{ atoms O}$

6.47 $0.20 \text{ moles } Cl_2 \times \frac{2 \text{ moles Cl atoms}}{1 \text{ mole } Cl_2} = 0.40 \text{ moles Cl atoms}$

6.49 $1432 \text{ moles } Na_2S_2O_4 \times \frac{2 \text{ moles S atoms}}{1 \text{ mole } Na_2S_2O_4} = 2864 \text{ moles S atoms}$

6.51 $22{,}000 \text{ atoms Ti} \times \frac{4 \text{ atoms Cl}}{1 \text{ atom Ti}} = 88{,}000 \text{ atoms Cl}$

6.53 $12{,}300 \text{ moles } Al_2S_3 \times \dfrac{3 \text{ moles S}}{1 \text{ mole } Al_2S_3} = 36{,}900 \text{ moles S}$

6.55 We know the alloy consists of 6% copper, 27% silver, 26% mercury—this leaves 41% cobalt.

$$0.376 \text{ g alloy} \times \frac{6 \text{ g Cu}}{100 \text{ g alloy}} = 0.0226 \text{ g Cu}$$

$$0.376 \text{ g alloy} \times \frac{27 \text{ g Ag}}{100 \text{ g alloy}} = 0.102 \text{ g Ag}$$

$$0.376 \text{ g alloy} \times \frac{26 \text{ g Hg}}{100 \text{ g alloy}} = 0.0978 \text{ g Hg}$$

$$0.376 \text{ g alloy} \times \frac{41 \text{ g Co}}{100 \text{ g alloy}} = 0.154 \text{ g Co}$$

6.57 $0.020 \text{ L} \times \dfrac{8920 \text{ g}}{\text{L}} = 178.4 \text{ g} \approx 180 \text{ g}$

6.59 $95.0 \text{ g cm}^3 \times \dfrac{0.00278 \text{ g}}{\text{cm}^3} = 0.264 \text{ g}$

6.61 $d = \dfrac{0.532 \text{ g}}{0.653 \text{ cm}^3} = 0.815 \text{ g/cm}^3$

6.63

a) $\dfrac{2.14 \text{ g}}{27.6 \text{ cm}^3} = \dfrac{17.2 \text{ g}}{? \text{ cm}^3}$

Answer: 22.2 cm^3

b) $d = \dfrac{21.4 \text{ g}}{27.6 \text{ cm}^3} = 0.775 \text{ g/cm}^3$

This matches substance B with $d = 0.780 \text{ g/cm}^3$.

d) Substance C has $d = 0.841 \text{ g/cm}^3$, hence it will float on liquids with a higher density and it will sink in liquids with a lower density.

Substance	Density (g/cm³)	Float?	Sink?
Water	1.001	✔	
Ethyl alcohol	0.791		✔
Benzene	0.877	✔	

6.65 $5000.0 \text{ cm}^3 \text{ sol} \times \dfrac{1.22 \text{ g sol}}{\text{cm}^3 \text{ sol}} \times \dfrac{34.6 \text{ g NaCl}}{100 \text{ g sol}} = 2110 \text{ g NaCl}$

6.67 $125.0 \text{ g sol} \times \frac{8.6 \text{ g Na}}{100 \text{ g sol}} = 11 \text{ g Na}$

$125.0 \text{ g sol} \times \frac{10.5 \text{ g glucose}}{100 \text{ g sol}} = 13.1 \text{ g glucose}$

$125.0 \text{ g sol} \times \frac{4.8 \text{ g K}}{100 \text{ g sol}} = 6.0 \text{ g K}$

6.69 $\frac{0.000068 \text{ g coloring agent}}{50 \text{ g shampoo}} = \frac{? \text{ g}}{10^6 \text{ g}}$

Answer: 1.36 ppm ~ 1.4 ppm (Answer should be expressed in two significant figures.)

6.71 $4000.0 \text{ g sample} \times \frac{6 \text{ g Pb}}{10^6 \text{ g sample}} = 0.024 \text{ g Pb} \approx 0.02 \text{ g Pb}$

Answer should be expressed with one significant figure.

6.73 mass = 83.18 g − 78.30 g = 4.88 g

volume = $28.30 \text{ cm}^3 - 25.05 \text{ cm}^3 = 3.25 \text{ cm}^3$

$d = \frac{4.88 \text{ g}}{3.25 \text{ cm}^3} = 1.50 \text{ g/cm}^3$

CHAPTER 7

7.1 It would be better to report "guessed" numbers than imprecise ones because imprecise numbers may be wrong, but could be taken for correct. By the nature of a "guessed" number, it is understood to have uncertainty in its value.

7.3 We count significant figures in a measured number to indicate the certainty of the measurement.

7.5 Measurements are made with devices (e.g., rulers, balances, cylinders) that are calibrated to a particular degree. Due to the calibration of these devices, there is uncertainty in the measurements made with them.

7.7 Whether you are multiplying or dividing, the answer should have the same number of significant figures as the least number of significant figures used in the entire mathematical operation. For example, if you were to multiply 1.23×4.5678, the answer would be reported to three significant figures (5.62).

7.9 Remember that nonzeros are significant. Trailing zeros are significant with a decimal point. Zeros in between nonzero digits are significant.

a) 3

b) 4

c) 1 (zeros are before nonzero)

d) 2

e) 2 (trailing zeros don't count, there's no decimal)

7.11 Let $a = 3$

$a^2 + a^2 = 3^2 + 3^2 = 9 + 9 = 18$

$(a^2)^2 = (3^2)^2 = (9)^2 = 81$

$a^4 = (3)^4 = 81$

So $a^2 + a^2 \neq (a^2)^2$

Now, $a^2 + a^2 = 3^2 + 3^2 = 18$

$2a^2 = 2(3)^2 = 2(9) = 18$

So, $a^2 + a^2 = 2a^2$

7.13 Scientific Notation is a way to represent numbers that are either very large or very small. It is written as the product of a number between 1 and 10 and raised to a power that may be positive or negative.

7.15 See Table 7-7.

a) 1×10^9 g = 1 Gg

b) 1×10^{-1} g = 1 dg

c) 1×10^3 = 1 kg

d) 1×10^{-3} g = 1 mg

e) 1×10^{-6} liters = 1 μL

7.17 Trailing zeros are significant only with a decimal point shown in the value.

a) one

b) two

c) five

7.19 a) −1

b) −1

7.21 $\frac{3x^{-2}}{4x^3} = (3x^{-2})(1/4x^{-3}) = 3/4x^{-5}$

7.23 $\frac{5x^{-2}y^{-3}}{3z^{-1}} = \frac{5z}{3x^2y^3}$

7.25 Remember, in scientific notation only the digits in the number that is multiplied by a power of 10 is significant. That number must be between one and 10.

a) 4.65×10^4

b) 3×10^{-4}

c) 3.61×10^6

d) 4.17089×10^1

e) 2.198×10^3

7.27 a) $1.63 \times 10^{-4} = 0.000163$

b) $2.317 \times 10^5 = 231{,}700$

c) $7.3 \times 10^6 = 7{,}300{,}000$

7.29 Any unit containing m is length, g is mass, and L is volume.

Measurement	Length, volume, or mass?	Number of significant figures
17.25 cm	Length	4
2.83 µg	Mass	3
1.0004×10^2 mL	Volume	5
32.400 kg	Mass	5
0.00068 pm	Length	2
1.48×10^{-2} dL	Volume	3
5.230×10^2	Volume	4

7.31 $d = \frac{1.50 \text{ g}}{1 \text{ mL}}$

$$\text{mass} = 25 \text{ mL} \times \frac{1.50 \text{ g}}{1 \text{ mL}} = 38 \text{ g}$$

7.33 We need to convert mL to L and seconds to minutes.

$$\frac{25.0\ \cancel{mL}}{\cancel{sec}} \times \frac{10^{-3}\ L}{1\ \cancel{mL}} \times \frac{60\ \cancel{sec}}{1\ min} = 1.5\ L/min$$

Answer should be reported with three significant figures; therefore, the answer is 1.50 L/min.

7.35 $$\frac{(3.64 \times 10^{81})}{(1.75 \times 10^{-72})} \cdot \frac{(8.85 \times 10^{-7})}{(6.94 \times 10^{135})} = \frac{(3.64) \times (8.85)}{(1.75) \times (6.94)} \times 10^{(81-7+72-135)} = 2.65 \times 10^{11}$$

7.37 $$1.07\ \cancel{Mm} \times \frac{10^6\ \cancel{m}}{1\ \cancel{Mm}} \times \frac{1\ km}{10^3\ \cancel{m}} = 1.07 \times 10^3\ km$$

7.39 $$7.53\ \cancel{GL} \times \frac{10^9\ \cancel{L}}{1\ \cancel{GL}} \times \frac{1\ \mu L}{10^{-6}\ \cancel{L}} = 7.53 \times 10^{15}\ \mu L$$

7.41 $$2.03 \times 10^4\ \cancel{mm} \times \frac{10^{-3}\ \cancel{m}}{1\ \cancel{mm}} \times \frac{1\ pm}{10^{-12}\ \cancel{m}} = 2.03 \times 10^5\ pm$$

7.43 $$1.89 \times -10^3\ mm \times \frac{10^6\ L}{1\ ML} \times \frac{1\ dL}{10^{-1}\ L} = 1.89 \times 10^4\ dL$$

7.45 a) $$10.52\ mL \times \frac{10^{-3}\ L}{1\ mL} = 1.052 \times 10^{-2}\ L$$

b) $$10.52\ L \times \frac{1\ mL}{10^{-3}\ L} = 1.50 \times 10^3\ mL$$

c) $$18.3\ cm^3 \times \frac{1\ mL}{1\ cm^3} = 18.3\ mL$$

d) $$125.0\ \cancel{cm^3} \times \frac{1\ \cancel{mL}}{1\ \cancel{cm^3}} \times \frac{10^{-3}\ L}{1\ \cancel{mL}} = 0.1250\ mL$$

e) $$0.1095\ g \times \frac{1\ mg}{10^{-3}\ g} = 109.5\ mg$$

f) $$65.2\ Kg \times \frac{10^3\ g}{1\ kg} = 6.52 \times 10^4\ g$$

g) $$5837\ g \times \frac{1\ kg}{10^3\ g} = 5.837\ kg$$

7.47 $$6.022 \times 10^{23}\ \text{atoms O} \times \frac{1\ \text{molecule } CO_2}{2\ \text{atoms O}} \times \frac{1\ \text{molecule } CO_2}{6.022 \times 10^{23}\ \text{molecules } CO_2}$$

$$= 0.5000\ \text{mole } CO_2$$

7.49 $34.5 \text{ g} \times \frac{1 \text{ cm}^3}{0.740 \text{ g}} = 46.6 \text{ cm}^3$

7.51 a) Adding the values and dividing by the number of measurements gives us:

$$\frac{0.08653}{7} = 0.01236 \text{ L}$$

b) The range is 0.01225 L to 0.01250 L.

c) $0.01236 \text{ L} \times \frac{125 \text{ g}}{\text{L}} = 1.54 \text{ g}$

7.53 $7x^{-4} \cdot 6x^{-3} = 42x^{-7}$

7.55 $\frac{15.0 \text{ g}}{22.4 \text{ mL}} = \frac{50.0 \text{ g}}{\text{X mL}} = 74.7 \text{ mL}$

7.57 First, we calculate the atoms of Pb in one glass of water (120 g) using the ratio for ppm: 26.80 g Pb/1,000,000 g H_2O.

$$\frac{26.80 \text{ g Pb}}{1 \times 10^6 \text{g } H_2O} = \frac{x \text{ g Pb}}{120\text{g } H_2O} \quad x = 0.003216 \text{ g Pb}$$

Second, we calculate moles Pb in this gram amount by dividing by molar mass. Finally, we calculate the number of Pb atoms contained in this mol amount, using Avogadro's number. These two calculations can be done consecutively.

$$(0.003216 \text{ g Pb}) \times \frac{1 \text{ mol Pb}}{207.2 \text{ g Pb}} \times \frac{6.023 \times 10^{23} \text{ atoms Pb}}{1 \text{ mol Pb}} = 9.374 \times 10^{18} \text{ atoms Pb}$$

7.59 $6.022 \times 10^{22} \text{ molecules} \times \frac{1 \text{ mole}}{6.022 \times 10^{23} \text{ molecules}} = 0.1000 \text{ mole}$

7.61 a) 4
b) 4
c) 7
d) 3
e) 5
f) 2

7.63 a) 8.73×10^{-7}
b) 6.00835×10^5
c) 5.26×10^9
d) 9.14×10^{-6}
e) 8.904×10^0 ($10^0 = 1$)

7.65 a) $250.0\ \text{mL} \times \dfrac{10^{-3}\ \text{L}}{1\ \text{mL}} = 0.2500\ \text{L}$

b) $15.0\ \text{L} \times \dfrac{1\ \text{mL}}{10^{-3}\ \text{L}} = 1.50 \times 10^4\ \text{mL}$

c) $25.95\ \text{cm}^3 \times \dfrac{1\ \text{mL}}{1\ \text{cm}^3} = 25.95\ \text{cm}^3$

d) $9.87\ \text{cm}^3 \times \dfrac{10^{-3}\ \text{L}}{1\ \text{cm}^3} = 9.87 \times 10^{-3}\ \text{L}$

e) $1.68 \times 10^5\ \mu\text{g} \times \dfrac{10^{-6}\ \text{g}}{1\ \mu\text{g}} = 0.168\ \text{g}$

d) $4.975\ \text{g} \times \dfrac{1\ \mu\text{g}}{10^{-6}\ \text{g}} = 4.975 \times 10^6\ \mu\text{g}$

e) $4.027 \times 10^{19}\ \cancel{\text{mg}} \times \dfrac{10^{-3}\ \cancel{\text{g}}}{1\ \cancel{\text{mg}}} \times \dfrac{10\ \text{kg}}{10^3\ \cancel{\text{g}}} = 4.027 \times 10^{13}\ \text{kg}$

CHAPTER 8

8.1

Conversion	**Ratio needed in calculation**
Grams H_3PO_4 → moles H_3PO_4	$\frac{1 \text{ mole } H_3PO_4}{97.994 \text{ g}}$
Number H atoms → molecules H_3PO_4	$\frac{1 \text{ molecule } H_3PO_4}{3 \text{ atoms H}}$
Molecules H_3PO_4 → number of O atoms	$\frac{4 \text{ atoms of O}}{1 \text{ molecule } H_3PO_4}$
Number of H atoms → moles $C_6H_{12}O_6$ molecules	$\frac{1 \text{ molecule } C_6H_{12}O_6}{12 \text{ atoms of H}}$ $\frac{1 \text{ mole } C_6H_{12}O_6}{6.022 \times 10^{23} \text{ molecules}}$
Moles P atoms → grams P	$\frac{30.974 \text{ g P}}{1 \text{ mole P}}$

8.3 $266 \text{ g } (NH_4)_3 PO_4 \times \frac{1 \text{ mole } (NH_4)_3 PO_4}{150.179 \text{ g } (NH_4)_3 PO_4} = 1.998 \text{ moles } (NH_4)_3 PO_4 \approx 2.00 \text{ moles}$

8.5 $0.0123 \text{ g } NO_2 \times \frac{1 \text{ mole } NO_2}{46.005 \text{ g } NO_2} = 2.67 \times 10^{-4} \text{ mole } NO_2$

8.7 $3.25 \text{ moles } KNO_3 \times \frac{101.102 \text{ g } KNO_3}{1 \text{ mole } KNO_3} = 328.58 \text{ g } KNO_3 \approx 329 \text{ g } KNO_3$

8.9 $4.95 \text{ moles } N_2O_5 \times \frac{108.009 \text{ g } N_2O_5}{1 \text{ mole } N_2O_5} = 535 \text{ g } N_2O_5$

8.11 a) $0.0010 \text{ mol } SF_4 \times \frac{108.058 \text{ g}}{1 \text{ mol}} = 0.11 \text{ g } SF_4$

b) $23.4 \text{ mol } CaCO_3 \times \frac{100.086 \text{ g}}{1 \text{ mol}} = 2340 \text{ g } CaCO_3$

8.13 a) $24.4 \text{ g } Ba(C_2H_3O_2)_2 \times \frac{1 \text{ mole } Ba(C_2H_3O_2)_2}{255.415 \text{ g}} \times \frac{4 \text{ mol C}}{1 \text{ mole } Ba(C_2H_3O_2)_2} = 0.382 \text{ mol C}$

b) $264.2 \text{ mol } (NH_4)_3 PO_4 \times \dfrac{12 \text{ mol H}}{1 \text{ mol } (NH_4)_3 PO_4} = 3170. \text{ mol H or } 3.170 \times 10^3 \text{ mol H}$

We need to show four significant figures. Without a decimal, there are only three significant figures. Scientific notation is the best way to express answers in situations where the answer ends with a zero.

8.15 a) $1 \text{ mol } \cancel{KBr} \times \dfrac{1 \cancel{\text{ mol K}}}{1 \text{ mol } \cancel{KBr}} \times \dfrac{39.098 \text{ g K}}{1 \cancel{\text{ mol K}}} = 39.098 \text{ g K}$

b) $1 \text{ mol } \cancel{Mn_2O_3} \times \dfrac{2 \text{ mol } \cancel{Mn}}{1 \text{ mol } \cancel{Mn_2O_3}} \times \dfrac{54.938 \text{ g Mn}}{1 \text{ mol } \cancel{Mn}} = 109.876 \text{ g Mn}$

c) $1 \text{ mol } \cancel{Fe_3O_4} \times \dfrac{3 \text{ mol } \cancel{Fe}}{1 \text{ mol } \cancel{Fe_3O_4}} \times \dfrac{55.847 \text{ g Fe}}{1 \cancel{\text{ mol Fe}}} = 167.541 \text{ g Fe}$

8.17 $0.4318 \text{ g } \cancel{FeCl_2} \times \dfrac{1 \text{ mol } FeCl_2}{126.753 \text{ g } \cancel{FeCl_2}} \times \dfrac{1 \text{ mol Fe}}{1 \text{ mol } FeCl_2} \times \dfrac{55.847 \text{ g Fe}}{1 \text{ mol Fe}} = 0.1902 \text{ g Fe}$

$0.4318 \text{ g } \cancel{FeCl_2} \times \dfrac{1 \text{ mol } FeCl_2}{126.753 \text{ g } \cancel{FeCl_2}} \times \dfrac{2 \text{ mol Cl}}{1 \text{ mol } FeCl_2} \times \dfrac{35.453 \text{ g Cl}}{1 \text{ mol Cl}} = 0.2416 \text{ g Cl}$

8.19 $0.025 \cancel{\text{L}} \times \dfrac{995 \cancel{\text{ g } H_2O}}{1 \cancel{\text{ L } H_2O}} \times \dfrac{1 \text{ mol } H_2O}{18.015 \cancel{\text{ g } H_2O}} = 1.4 \text{ mol } H_2O$

8.21 Gases:

a) $0.0010 \text{ L} \times \dfrac{0.084 \text{ g}}{\text{L}} \times \dfrac{1 \text{ mol } H_2}{2.016 \text{ g}} = 0.000042 \text{ mol } H_2$

b) $0.0010 \text{ L} \times \dfrac{2.95 \text{ g}}{\text{L}} \times \dfrac{1 \text{ mol } Cl_2}{70.906 \text{ g}} = 0.000042 \text{ mol } Cl_2$

c) $0.0010 \text{ L} \times \dfrac{0.666 \text{ g}}{\text{L}} \times \dfrac{1 \text{ mol } CH_4}{16.043 \text{ g}} = 0.000042 \text{ mol } CH_4$

d) $0.0010 \text{ L} \times \dfrac{5.59 \text{ g}}{\text{L}} \times \dfrac{1 \text{ mol Xe}}{131.29 \text{ g}} = 0.000043 \text{ mol Xe}$

Liquids:

e) $0.0010 \text{ L} \times \dfrac{1260 \text{ g}}{\text{L}} \times \dfrac{1 \text{ mol } C_3H_8O}{92.094 \text{ g}} = 0.014 \text{ mol } C_3H_8O$

f) $0.0010 \cancel{\text{L}} \times \dfrac{995 \text{ g}}{\cancel{\text{L}}} \times \dfrac{1 \text{ mol } H_2O}{18.015 \text{ g}} = 0.055 \text{ mol } H_2O$

g) $0.0010 \text{ L} \times \dfrac{1460 \text{ g}}{\text{L}} \times \dfrac{1 \text{ mol } C_2H_5Br}{108.966 \text{ g}} = 0.013 \text{ mol } C_2H_5Br$

h) $0.0010 \cancel{\text{L}} \times \dfrac{730 \text{ g}}{\cancel{\text{L}}} \times \dfrac{1 \text{ mol } C_{10}H_{22}}{142.286 \text{ g}} = 0.0051 \text{ mol } C_{10}H_{22}$

i) $0.0010\,\text{L} \times \dfrac{630\,\text{g}}{\text{L}} \times \dfrac{1\text{ mol } C_5H_{12}}{72.151\,\text{g}} = 0.0087\text{ mol } C_5H_{22}$

Metals:

j) $0.0010\,\text{L} \times \dfrac{11{,}400\,\text{g}}{\text{L}} \times \dfrac{1\text{ mol Pb}}{207.2\text{ g Pb}} = 0.055\text{ mol Pb}$

k) $0.0010\,\text{L} \times \dfrac{7860\,\text{g}}{\text{L}} \times \dfrac{1\text{ mol Fe}}{55.847\,\text{g}} = 0.14\text{ mol Fe}$

l) $0.0010\,\text{L} \times \dfrac{2700\,\text{g}}{\text{L}} \times \dfrac{1\text{ mol Al}}{26.982\,\text{g}} = 0.10\text{ mol Al}$

m) $0.0010\,\text{L} \times \dfrac{19{,}800\,\text{g}}{\text{L}} \times \dfrac{1\text{ mol Pu}}{244\,\text{g}} = 0.081\text{ mol Pu}$

Gases:
The greater the molar mass, the higher the density. Equal volumes of gases contain the same number of moles (at constant temperature and pressure).

Liquids:
The density of liquids don't seem to follow a trend.

Metals:
The greater the molar mass, the higher the density.

8.23 a)

$$\frac{2\text{ mol C} \times \dfrac{12.011\text{ g C}}{\text{mol C}}}{1\text{ mol } C_2H_4O_2 \times \dfrac{60.052\text{ g } C_2H_4O_2}{\text{mol } C_2H_4O_2}} \times 100 = 40.002\%\text{ C}$$

b)

$$\frac{6\text{ mol C} \times \dfrac{12.011\text{ g C}}{\text{mol C}}}{1\text{ mol } C_6H_{12}O_6 \times \dfrac{180.156\text{ g } C_6H_{12}O_6}{\text{mol } C_6H_{12}O_6}} \times 100 = 40.002\%\text{ C}$$

c)

$$\frac{5\text{ mol C} \times \dfrac{12.011\text{ g C}}{\text{mol C}}}{1\text{ mol } C_5H_{10}O_5 \times \dfrac{150.130\text{ g } C_5H_{10}O_5}{\text{mol } C_5H_{10}O_5}} \times 100 = 40.002\%\text{ C}$$

d)

$$\frac{3\text{ mol C} \times \dfrac{12.011\text{ g C}}{\text{mol C}}}{1\text{ mol } C_6H_{12} \times \dfrac{42.081\text{ g } C_3H_6}{\text{mol } C_3H_6}} \times 100 = 85.628\%\text{ C}$$

e)

$$\frac{6\text{ mol C} \times \dfrac{12.011\text{ g C}}{\text{mol C}}}{1\text{ mol } C_6H_{12} \times \dfrac{84.162\text{ g } C_6H_{12}}{\text{mol } C_6H_{12}}} \times 100 = 85.628\%\text{ C}$$

f) $$\frac{2 \text{ mol C} \times \frac{12.011 \text{ g C}}{\text{mol C}}}{1 \text{ mol } C_2H_4 \times \frac{28.054 \text{ g } C_2H_4}{\text{mol } C_2H_4}} \times 100 = 85.628\% \text{ C}$$

The simplest formula for a, b, and c is CH_2O. We see that these three compounds have the same mass percentage of carbon.

The simplest formula for d, e, and f is CH_2 with the same mass percentage as carbon.

8.25 This requires you to determine the mass percentage of each element in the compound. To do this, you will have a separate calculation for each element.

a) KOH: molar mass $= \frac{56.105 \text{ g KOH}}{\text{mol KOH}}$

mass % K: $$\frac{1 \text{ mol K} \times \frac{39.098 \text{ g K}}{\text{mol K}}}{1 \text{ mol KOH} \times \frac{56.105 \text{ g KOH}}{\text{mol KOH}}} \times 100 = 69.687\% \text{ K}$$

mass % O: $$\frac{1 \text{ mol O} \times \frac{15.999 \text{ g O}}{\text{mol O}}}{1 \text{ mol KOH} \times \frac{56.105 \text{ g KOH}}{\text{mol KOH}}} \times 100 = 28.516\% \text{ O}$$

mass % H: $$\frac{1 \text{ mol H} \times \frac{1.008 \text{ g H}}{\text{mol H}}}{1 \text{ mol KOH} \times \frac{56.105 \text{ g KOH}}{\text{mol KOH}}} \times 100 = 1.796\% \text{ H}$$

b) RbBr: molar mass $= \frac{165.372 \text{ g}}{\text{mol}}$

mass % Rb: $$\frac{1 \text{ mol Rb} \times \frac{85.468 \text{ g Rb}}{\text{mol Rb}}}{1 \text{ mol RbBr} \times \frac{165.372 \text{ g RbBr}}{\text{mol RbBr}}} \times 100 = 51.682\% \text{ Rb}$$

mass % Br: $$\frac{1 \text{ mol Br} \times \frac{79.904 \text{ g Br}}{\text{mol Br}}}{1 \text{ mol RbBr} \times \frac{165.372 \text{ g RbBr}}{\text{mol RbBr}}} \times 100 = 43.318\% \text{ Br}$$

c) $LaCl_3$: molar mass $= \frac{245.265 \text{ g}}{\text{mol}}$

mass % La: $$\frac{1 \text{ mol La} \times \frac{139.906 \text{ g La}}{\text{mol La}}}{1 \text{ mol } LaCl_3 \times \frac{245.265 \text{ g } LaCl_3}{\text{mol } LaCl_3}} \times 100 = 56.635\% \text{ La}$$

mass % Cl: $$\frac{3 \text{ mol Cl} \times \frac{35.453 \text{ g Cl}}{\text{mol Cl}}}{1 \text{ mol } LaCl_3 \times \frac{245.265 \text{ g } LaCl_3}{\text{mol } LaCl_3}} \times 100 = 48.365\% \text{ Cl}$$

d) $LuCl_3$: molar mass $= \dfrac{281.326\ g}{mol}$

mass % Lu: $$\frac{1\ mol\ Lu \times \frac{174.967\ g\ Lu}{mol\ Lu}}{1\ mol\ LuCl_3 \times \frac{281.326\ g\ LuCl_3}{mol\ LuCl_3}} \times 100 = 62.194\%\ Lu$$

mass % Cl: $$\frac{3\ mol\ Cl \times \frac{35.453\ g\ Cl}{mol\ Cl}}{1\ mol\ LuCl_3 \times \frac{281.326}{mol\ LuCl_3}} \times 100 = 37.806\%\ Cl$$

Because of significant figures, the total of mass percentages may not be exactly one hundred percent.

8.27 Using the mass percentage of copper in each ore, you can calculate the mass of copper in each.

a) $$100\ g\ Cu_2S \times \frac{mol\ Cu_2S}{159.158\ g\ Cu_2S} \times \frac{2\ mol\ Cu}{1\ mol\ Cu_2S} \times \frac{63.546\ g\ Cu}{mol\ Cu} = 79.85\ g\ Cu \approx 80\ g\ Cu$$

b) $$100\ g\ Cu_2CO_3(OH)_2 \times \frac{mol\ Cu_2CO_3(OH)_2}{210.069\ Cu_2CO_3(OH)_2} \times \frac{2\ mol\ Cu}{1\ mol\ Cu_2CO_3(OH)_2} \times \frac{63.546\ g\ Cu}{mol\ Cu}$$

$= 60.500\ g\ Cu \approx 60\ g\ Cu$

c) $$100\ g\ CuFeS_2 \times \frac{mol\ CuFeS_2}{183.523\ g\ CuFeS_2} \times \frac{1\ mol\ Cu}{1\ mol\ CuFeS_2} \times \frac{63.546\ g\ Cu}{mol\ Cu}$$

$= 34.63\ g\ Cu \approx 30\ g\ Cu$

Answer can only have one significant figure!

8.29 a) mass % K: $$\frac{2\ mol\ K \times \frac{39.098\ g\ K}{mol\ K}}{1\ mol\ K_2SO_4 \times \frac{174.258\ g\ K_2SO_4}{mol\ K_2SO_4}} \times 100 = 44.874\%\ K$$

mass % S: $$\frac{1\ mol\ S \times \frac{32.066\ g\ S}{mol\ S}}{1\ mol\ K_2SO_4 \times \frac{174.258\ g\ K_2SO_4}{mol\ K_2SO_4}} \times 100 = 18.401\%\ S$$

mass % O: $$\frac{4\ mol\ O \times \frac{15.999\ g\ O}{mol\ O}}{1\ mol\ K_2SO_4 \times \frac{174.258\ g\ K_2SO_4}{mol\ K_2SO_4}} \times 100 = 36.725\%\ O$$

b) mass % Fe: $$\frac{1 \text{ mol Fe} \times \frac{55.847 \text{ g Fe}}{\text{mol Fe}}}{1 \text{ mol Fe}(C_5H_5)_2 \times \frac{186.037 \text{ g Fe}(C_5H_5)_2}{\text{mol Fe}(C_5H_5)_2}} \times 100 = 30.019\% \text{ Fe}$$

mass % C: $$\frac{10 \text{ mol C} \times \frac{12.011 \text{ g C}}{\text{mol C}}}{1 \text{ mol Fe}(C_5H_5)_2 \times \frac{186.037 \text{ g Fe}(C_5H_5)_2}{\text{mol Fe}(C_5H_5)_2}} \times 100 = 64.562\% \text{ C}$$

mass % H: $$\frac{10 \text{ mol H} \times \frac{1.008 \text{ g H}}{\text{mol H}}}{1 \text{ mol Fe}(C_5H_5)_2 \times \frac{186.037 \text{ g Fe}(C_5H_5)_2}{\text{mol Fe}(C_5H_5)_2}} \times 100 = 5.418\% \text{ H}$$

c) mass % Ba: $$\frac{1 \text{ mol Ba} \times \frac{137.327 \text{ g Ba}}{\text{mol Ba}}}{1 \text{ mol Ba}(C_2H_3O_2)_2 \times \frac{255.415 \text{ g Ba}(C_2H_3O_2)_2}{\text{mol Ba}(C_2H_3O_2)_2}} \times 100 = 53.766\% \text{ Ba}$$

mass % C: $$\frac{4 \text{ mol C} \times \frac{12.011 \text{ g C}}{\text{mol C}}}{1 \text{ mol Ba}(C_2H_3O_2)_2 \times \frac{255.415 \text{ g Ba}(C_2H_3O_2)_2}{\text{mol Ba}(C_2H_3O_2)_2}} \times 100 = 18.810\% \text{ C}$$

mass % H: $$\frac{6 \text{ mol H} \times \frac{1.008 \text{ g H}}{\text{mol H}}}{1 \text{ mol Ba}(C_2H_3O_2)_2 \times \frac{255.415 \text{ g Ba}(C_2H_3O_2)_2}{\text{mol Ba}(C_2H_3O_2)_2}} \times 100 = 2.368\% \text{ H}$$

mass % O: $$\frac{4 \text{ mol O} \times \frac{15.999 \text{ g O}}{\text{mol O}}}{1 \text{ mol Ba}(C_2H_3O_2)_2 \times \frac{255.415 \text{ g Ba}(C_2H_3O_2)_2}{\text{mol Ba}(C_2H_3O_2)_2}} \times 100 = 25.056\% \text{ O}$$

8.31 a) mass % Na: $$\frac{1 \text{ mol Na} \times \frac{22.990 \text{ g Na}}{\text{mol Na}}}{1 \text{ mol NaCl} \times \frac{58.443 \text{ g NaCl}}{\text{mol NaCl}}} \times 100 = 39.337\% \text{ Na}$$

b) mass % Pb: $$\frac{1 \text{ mol Pb} \times \frac{207.2 \text{ g Pb}}{\text{mol Pb}}}{1 \text{ mol PbCl}_2 \times \frac{278.1 \text{ g PbCl}_2}{\text{mol PbCl}_2}} \times 100 = 74.50\% \text{ Pb}$$

c) mass % U: $$\frac{1 \text{ mol U} \times \frac{238.029 \text{ g U}}{\text{mol U}}}{1 \text{ mol UCl}_2 \times \frac{379.841 \text{ g UCl}_4}{\text{mol UCl}_4}} \times 100 = 62.6654\% \text{ U}$$

8.33 Let's start with Li_2CO_3.

$$\text{mass \% Li: } \frac{2 \text{ mol Li} \times \frac{6.941 \text{ g Li}}{\text{mol Li}}}{1 \text{ mol } Li_2CO_3 \times \frac{73.890 \text{ g } Li_2CO_3}{\text{mol } Li_2CO_3}} \times 100 = 18.79\% \text{ Li}$$

This matches the mass percentage given in the problem; therefore, M is lithium.

In Problem 12 the molar mass *range* is given, not an exact molar mass. Here the mass percentages are given to two decimals, hence more accurate. Mass percentages are more useful than molar masses for determining the identity of an unknown metal.

8.35
- Empirical is an adjective meaning based on or characterized by observation and experiment rather than theory.
- Empirical formula is a formula derived from experimentation and measurement.

8.37 To find the empirical formula, we will get a ratio of the elements. This is done by dividing both by the smallest value.

$$\frac{0.25 \text{ mol C}}{0.25 \text{ mol compound}} = \frac{1 \text{ mol C}}{1 \text{ mol cmpd}}$$

$$\frac{0.625 \text{ mol H}}{0.25 \text{ mol cmpd}} = \frac{2.5 \text{ mol H}}{1 \text{ mol cmpd}}$$

Since we get a fraction for H, we must double both to get a whole number ratio. The empirical formula is C_2H_5.

8.39 Assuming a 100.0 gram sample, we have 28.03 g Na, 39.0 g O, 3.69 g H, and 29.3 g C. To get an empirical formula, we must convert the gram amounts of each element to moles; then, we can get a ratio by dividing by the smallest one.

$$28.03 \text{ g Na} \times \frac{0.1 \text{ mol Na}}{22.990 \text{ g Na}} = \frac{1.219 \text{ mol Na}}{1.219 \text{ mol cmpd}} = 1 \text{ Na}$$

$$39.0 \text{ g O} \times \frac{1 \text{ mol O}}{15.999 \text{ g O}} = \frac{2.44 \text{ mol O}}{1.219 \text{ mol cmpd}} = 2 \text{ O}$$

$$3.69 \text{ g H} \times \frac{1 \text{ mol H}}{1.008 \text{ g H}} = \frac{3.66 \text{ mol H}}{1.219 \text{ mol cmpd}} = 3 \text{ H}$$

$$29.3 \text{ g C} \times \frac{1 \text{ mol C}}{12.011 \text{ g C}} = \frac{2.44 \text{ mol C}}{1.219 \text{ mol cmpd}} = 2 \text{ C}$$

Empirical formula = $NaC_2H_3O_2$

8.41 Assuming 100.0 g sample:

$$35.9 \text{ g Al} \times \frac{1 \text{ mol Al}}{26.982 \text{ g Al}} = \frac{1.33 \text{ mol Al}}{1.33 \text{ mol cmpd}} = 1 \text{ Al}$$

$$64.1 \text{ g S} \times \frac{1 \text{ mol S}}{32.066 \text{ g S}} = \frac{2.00 \text{ mol S}}{1.33 \text{ mol cmpd}} = 1.5 \text{ S}$$

Since we got a fraction ratio (1: 1.5), we need to double both to get a whole number.

Empirical formula = Al_2S_3

8.43 Assuming 100.0 g sample:

$$79.96 \text{ g C} \times \frac{\text{mol C}}{12.011 \text{ g C}} = 6.657 \text{ mol C}$$

$$9.39 \text{ g H} \times \frac{\text{mol H}}{1.0079 \text{ g H}} = 9.32 \text{ mol H}$$

$$10.65 \text{ g O} \times \frac{\text{mol O}}{15.999 \text{ g O}} = 0.6657 \text{ mol O}$$

If there's 0.6657 mol substance, there's:

$$\frac{6.657 \text{ mol C}}{0.6657 \text{ mol substance}} = 10 \text{ C}$$

$$\frac{9.22 \text{ mol H}}{0.6657 \text{ mol substance}} = 14 \text{ H}$$

$$\frac{0.6657 \text{ mol O}}{0.6657 \text{ mol substance}} = 1 \text{ O}$$

The empirical formula is $C_{10}H_{14}O$.

8.45 The simplest formula CH_2 has mass 14.0268 g · mol^{-1}. This is half the molecular mass; hence, the molecular formula is double the empirical formula, C_2H_4.

8.47 Assuming 100.0 g sample:

$$(100 \text{ g} - 94.07 \text{ g O}) = 5.93 \text{ g H} \times \frac{\text{mol H}}{1.0079 \text{ g H}} \times \frac{5.88 \text{ mol H}}{5.88 \text{ mol substance}} \rightarrow 1\text{H}$$

$$94.07 \text{ g O} \times \frac{\text{mol O}}{15.999 \text{ g O}} \times \frac{5.88 \text{ mol O}}{5.88 \text{ mol substance}} \rightarrow 1\text{O}$$

The empirical formula is HO. The empirical mass is 17.00 g · mol^{-1}, half the molecular mass. The molecular formula is twice the empirical formula, H_2O_2.

8.49 Since we are given gram amounts of the two elements, we need not assume a 100 g sample. We can get moles of each element by dividing each mass by the respective elements' molar mass. We then continue, as we have done, by getting a molar ratio and the formula.

$$2.182 \text{ g P} \times \frac{\text{mol P}}{30.974 \text{ g P}} \times \frac{0.07045 \text{ mol P}}{0.07045 \text{ mol substance}} \rightarrow 1 \text{ P}$$

$$2.818 \text{ g O} \times \frac{\text{mol O}}{15.999 \text{ g O}} \times \frac{0.1761 \text{ mol O}}{0.07045 \text{ mol substance}} \rightarrow 2.5 \text{ O}$$

To get whole number ratios, we must double both values. The empirical formula is P_2O_5. The empirical mass is 141.943 g · mol^{-1}, half the molar mass. The molecular formula is P_4O_{10}.

8.51 In 100 g of compound:

$$13.88 \text{ g Li} \times \frac{1 \text{ mol Li}}{6.941 \text{ Li}} = 2.000 \text{ mol Li}$$

$$23.57 \text{ g C} \times \frac{1 \text{ mol C}}{12.011 \text{ g C}} = 1.9624 \text{ mol C}$$

$$(100 - (13.88 + 23.57)) \text{ g O} \times \frac{1 \text{ mol O}}{15.999 \text{ g O}} = 3.910 \text{ mol O}$$

The empirical formula is $LiCO_2$. The empirical molar mass is 1(6.941 g/mol) + 1(12.011 g/mol) + 2(15.999 g/mol) = 50.950 g/mol. This is half the molar mass, 101 g/mol. The molecular formula is $Li_2C_2O_4$, or lithium oxalate

8.53 The product contains tin and fluorine. Therefore, the mass of the product is equal to the mass of tin plus the mass of fluorine.

$$5.3 \text{ g Sn} \times \frac{\text{mol Sn}}{118.71 \text{ g Sn}} \times \frac{0.045 \text{ mol Sn}}{0.045 \text{ mol substance}} \rightarrow 1 \text{ Sn}$$

$$(8.7 \text{ g} - 5.3 \text{ g}) \text{ F} \times \frac{\text{mol F}}{18.998 \text{ g F}} \times \frac{0.18 \text{ mol F}}{0.045 \text{ mol substance}} \rightarrow 4 \text{ F}$$

The empirical formula is SnF_4.

8.55 a) PBr_3

$$\left(1 \text{ mol P} \times \frac{30.974 \text{ g P}}{\text{mol P}}\right) + \left(3 \text{ mol Br} \times \frac{79.904 \text{ g Br}}{\text{mol Br}}\right) = \frac{270.686 \text{ g PBr}_3}{\text{mol PBr}_3}$$

b) YCl_3

$$\left(1 \text{ mol Y} \times \frac{88.906 \text{ g Y}}{\text{mol Y}}\right) + \left(3 \text{ mol Cl} \times \frac{35.453 \text{ g Cl}}{\text{mol Cl}}\right) = \frac{195.265 \text{ g YCl}_3}{\text{mol YCl}_3}$$

c) $Ba(C_2H_3O_2)_2$

$$\left(1 \text{ mol Ba} \times \frac{137.327 \text{ g Ba}}{\text{mol Ba}}\right) + \left(4 \text{ mol C} \times \frac{12.011 \text{ g C}}{\text{mol C}}\right) + \left(6 \text{ mol H} \times \frac{1.008 \text{ g H}}{\text{mol H}}\right)$$

$$+ \left(4 \text{ mol O} \times \frac{15.999 \text{ g O}}{\text{mol O}}\right) = \frac{255.415 \text{ g Ba(C}_2\text{H}_3\text{O}_2)_2}{\text{mol Ba(C}_2\text{H}_3\text{O}_2)_2}$$

d) $(NH_4)_3PO_3$

$$\left(3 \text{ mol N} \times \frac{14.007 \text{ g N}}{\text{mol N}}\right) + \left(12 \text{ mol H} \times \frac{1.008 \text{ g H}}{\text{mol H}}\right) + \left(1 \text{ mol P} \times \frac{30.974 \text{ g P}}{\text{mol P}}\right)$$

$$+ \left(3 \text{ mol O} \times \frac{15.999 \text{ g O}}{\text{mol O}}\right) = \frac{133.088 \text{ g (NH}_4)_3\text{PO}_3}{\text{mol (NH}_4)_3\text{PO}_3}$$

e) $NaHSO_4$

$$\left(1 \text{ mol Na} \times \frac{22.990 \text{ g Na}}{\text{mol Na}}\right) + \left(1 \text{ mol H} \times \frac{1.008 \text{ g H}}{\text{mol H}}\right) + \left(1 \text{ mol S} \times \frac{32.066 \text{ g S}}{\text{mol S}}\right)$$

$$+ \left(4 \text{ mol O} \times \frac{15.999 \text{ g O}}{\text{mol O}}\right) = \frac{120.060 \text{ g NaHSO}_4}{\text{mol NaHSO}_4}$$

8.57 If we convert the volume to grams (using density), we can determine the number of moles using the molar mass. Remember that bromine is diatomic, Br_2.

$$250 \text{ cm}^3 \text{ Br}_2 \times \frac{3.119 \text{ g Br}_2}{\text{cm}^3 \text{ Br}_2} \times \frac{\text{mol Br}_2}{159.808 \text{ g Br}_2} = 4.9 \text{ mol Br}_2$$

8.59 a) $$1.00 \text{ g Os O}_4 \times \left[\frac{1 \text{ mol Os} \times \frac{190.2 \text{ g Os}}{\text{mol Os}}}{1 \text{ mol Os O}_4 \times \frac{254.2 \text{ g Os O}_4}{\text{mol Os O}_4}}\right] = 0.748 \text{ g Os}$$

b) $$1.00 \text{ g Ca(ClO}_4)_2 \times \left[\frac{1 \text{ mol Ca} \times \frac{40.078 \text{ g Ca}}{\text{mol Ca}}}{1 \text{ mol Ca(ClO}_4)_2 \times \frac{238.976 \text{ g Ca(ClO}_4)_2}{\text{mol Ca(ClO}_4)_2}}\right] = 0.168 \text{ g Ca}$$

8.61 $$2.33 \text{ mol ZnCl}_2 \cdot (\text{NH}_3)_4 \times \frac{1 \text{ mol Zn}}{1 \text{ mol ZnCl}_2 \cdot (\text{NH}_3)_4} \times \frac{65.39 \text{ g Zn}}{\text{mol Zn}} = 152 \text{ g Zn}$$

8.63 Empirical formulas must have the simplest ratios.

a) $C_3H_4Cl_2$: The empirical formula is CH_2Cl.

b) $Hg_2(C_2H_3O_2)_2$: The empirical formula is $HgC_2H_3O_2$.

c) $N_6H_{24}S_6$: The empirical formula is NH_4S.

8.65 We need to get a ratio.

$$\frac{1.40 \times 10^{-2} \text{ mol CaSO}_4}{1.40 \times 10^{-2} \text{ mol cmpd}} = 1 \text{ CaSO}_4$$

$$\frac{2.80 \times 10^{-2} \text{ mol H}_2\text{O}}{1.40 \times 10^{-2} \text{ mol cmpd}} = 2 \text{ H}_2\text{O}$$

The empirical formula is $CaSO_4 \cdot 2(H_2O)$.

8.67 Assume a 100.0 g sample.

$$14.87 \text{ g P} \times \frac{\text{mol P}}{30.974 \text{ g P}} \Rightarrow \frac{0.4801 \text{ mol P}}{0.4801 \text{ mol}} \Rightarrow 1 \text{ P}$$

$$85.13 \text{ g Cl} \times \frac{\text{mol Cl}}{35.453 \text{ g Cl}} \Rightarrow \frac{2.401 \text{ mol Cl}}{0.4801 \text{ mol}} \Rightarrow 5 \text{ Cl}$$

The empirical formula is PCl_5.

8.69 The formula NaO has a molar mass of 38.989 g. The molecule has a molar mass of 78 g. The empirical formula's mass is half the molar mass. The molecular formula is twice the empirical formula, or Na_2O_2.

8.71 See problems 8.67 and 8.69.

Assume a 100.0 g sample.

$$30.45 \text{ g N} \times \frac{\text{mol N}}{14.007 \text{ g N}} \Rightarrow \frac{2.174 \text{ mol N}}{2.174 \text{ mol}} \Rightarrow 1 \text{ N}$$

$$69.55 \text{ g O} \times \frac{\text{mol O}}{15.999 \text{ g O}} \Rightarrow \frac{4.347 \text{ mol O}}{2.174 \text{ mol}} \Rightarrow 2 \text{ O}$$

The empirical formula is NO_2. The empirical mass is 46.005 g mol^{-1}. This is half the molar mass. The molecular formula is N_2O_4.

8.73 a) $4.00 \text{ mol Ca(ClO}_3) \times \frac{206.978 \text{ g}}{1 \text{ mol}} = 828 \text{ g Ca(ClO}_3)_2$

b) $2100 \text{ mol NH}_4\text{H}_2\text{PO}_4 \times \frac{115.025 \text{ g}}{1 \text{ mol}} = 240{,}000 \text{ g NH}_4\text{H}_2\text{PO}_4$

8.75 a) $NaS_{0.5}O_2$: The ratios must be whole numbers. The empirical formula is Na_2SO_4.

b) $(NH_4)_2S_2O_8$: The empirical formula is NH_4SO_4.

CHAPTER 9

9.1 We saw in Chapter 4 that chemical reactions involve reactants with atoms in a particular arrangement or combination that become products. The products contain the same atoms as the reactants, the atoms simply have new arrangements. Therefore, the number of atoms does not change in going from reactants to products, they are merely recombining with each other.

9.3 We will need the ratios of the coefficients from the balanced equation.

a) $$126 \text{ molecules } C_4H_{10} \times \frac{8 \text{ molecules } CO_2}{2 \text{ molecules } C_4H_{10}} = 104 \text{ molecules } CO_2$$

b) $$26 \text{ molecules } O_2 \times \frac{8 \text{ molecules } CO_2}{13 \text{ molecules } O_2} = 16 \text{ molecules } CO_2$$

9.5 $4Al + 3O_2 \rightarrow 2Al_2O_3$

a) $$126 \text{ atoms Al} \times \frac{2 \text{ formula units } Al_2O_3}{4 \text{ atoms Al}} = 63 \text{ formula units } Al_2O_3$$

b) $$126 \text{ atoms O} \times \frac{1 \text{ molecule } O_2}{2 \text{ atoms O}} \times \frac{2 \text{ formula units } Al_2O_3}{3 \text{ molecules } O_2} = 42 \text{ formula units } Al_2O_3$$

c) In order to form one formula unit of Al_2O_3, two atoms of Al and three atoms of O are needed. There are more O atoms per formula unit than Al atoms. Although 126 atoms of Al would result in 63 formula units of Al_2O_3, we can get no more than 42 formula units of Al_2O_3 from 126 atoms of O. The number of O atoms limits the formation of the product.

d) $$28 \text{ formula units } Al_2O_3 \times \frac{3 \text{ molecules } O_2}{2 \text{ formula units } Al_2O_3} = 42 \text{ formula units } O_2$$

9.7 $2K + 2H_2O \rightarrow 2KOH + H_2$

The molar ratios are obtained from the coefficients in the balanced equation.

a) $$314 \text{ mol K} \times \frac{1 \text{ mol } H_2}{2 \text{ mol K}} = 157 \text{ mol } H_2$$

b) $$54 \text{ mol KOH} \times \frac{2 \text{ mol } H_2O}{2 \text{ mol KOH}} \times \frac{2 \text{ mol H}}{1 \text{ mol } H_2O} = 108 \text{ mol H}$$

c) $$150 \text{ mol } H_2 \times \frac{2 \text{ mol } H_2O}{1 \text{ mol } H_2} = 300 \text{ mol } H_2O$$

d) We need to calculate moles H produced in H_2 and moles H produced in KOH. The total amount of H will be the sum of the two numbers.

$$82 \text{ mol K} \times \frac{1 \text{ mol } H_2}{2 \text{ mol K}} \times \frac{2 \text{ mol K}}{1 \text{ mol } H_2} = 82 \text{ mol H}$$

$$82 \text{ mol K} \times \frac{2 \text{ mol KOH}}{2 \text{ mol K}} \times \frac{1 \text{ mol H}}{1 \text{ mol KOH}} = 82 \text{ mol H}$$

82 mol H (from H_2) + 82 mole H (from KOH) = 164 mol H

9.9 $CaCO_3 \rightarrow CaO + CO_2$

a) $$4520 \text{ mol } CaCO_3 \times \frac{1 \text{ mol CaO}}{1 \text{ mol } CaCO_3} = 4520 \text{ mol CaO}$$

9.11 $4Na_2CO_3 + Fe_3Br_8 \rightarrow 8NaBr + 4CO_2 + Fe_3O_4$

$$2450 \text{ mol NaBr} \times \frac{4 \text{ mol } Na_2CO_3}{8 \text{ mol NaBr}} = 1225 \text{ mol } Na_2CO_3$$

$$2450 \text{ mol NaBr} \times \frac{4 \text{ mol } Fe_3Br_8}{8 \text{ mol NaBr}} = 306 \text{ mol } Fe_3Br_8$$

9.13 $3Li + N_2 \rightarrow Li_3N$

a) $$0.304 \text{ mol Li} \times \frac{1 \text{ mol } Li_3N}{3 \text{ mol Li}} = 0.101 \text{ mol } Li_3N$$

b) $$3.04 \times 10^2 \text{ mol Li} \times \frac{1 \text{ mol } N_2}{3 \text{ mol Li}} = 1.01 \times 10^2 \ N_2$$

9.15 $2Al + 3S \rightarrow Al_2S_3$

a) $$25 \text{ mol Al} \times \frac{1 \text{ mol } Al_2S_3}{2 \text{ mol Al}} = 13 \text{ mol } Al_2S_3$$

b) $$15.5 \text{ mol S} \times \frac{1 \text{ mol } Al_2S_3}{3 \text{ mol S}} = 5.17 \text{ mol } Al_2S_3$$

9.17 $2KMnO_4 + 16\ HCl \rightarrow 2KCl + 2MnCl_2 + 5Cl_2 + 8H_2O$

$$1.05 \times 10^3 \text{ mol } KMnO_4 \times \frac{16 \text{ mol HCl}}{2 \text{ mol } KMnO_4} \times \frac{6.022 \times 10^{23} \text{ molecules HCl}}{1 \text{ mol HCl}}$$
$$= 5.06 \times 10^{27} \text{ g molecules HCl}$$

Here we have to remember the definition of a mole of molecules as 6.022×10^{23} molecules.

9.19 $2KMnO_4 + 16\ HCl \rightarrow 2KCl + 2MnCl_2 + 5Cl_2 + 8H_2O$

$$216.5 \cancel{\text{ g } Cl_2} \times \frac{\cancel{\text{mol } Cl_2}}{70.906 \cancel{\text{ g } Cl_2}} \times \frac{8 \cancel{\text{ mol } H_2O}}{5 \cancel{\text{ mol } Cl_2}} \times \frac{18.015 \text{ g } H_2O}{1 \cancel{\text{ mol } H_2O}} = 88.01 \text{ g } H_2O$$

Again, we need to use the molar masses to convert from grams to moles for Cl_2 and moles to grams for H_2O.

9.21 $$1.55 \text{ mol HNCO} \times \frac{4 \text{ mol } CO_2}{4 \text{ mol HNCO}} = 1.55 \text{ mol } CO_2$$

9.23 $$11.3 \text{ mol HNCO} \times \frac{6 \text{ mol NO}}{4 \text{ mol HNCO}} = 17.0 \text{ mol NO}$$

9.25 Here we must use the molar mass of HNCO and Avogadro's number to get molecules.

$$525 \text{ g HNCO} \times \frac{\text{mol HNCO}}{43.025 \text{ g HNCO}} \times \frac{5 \text{ mol N}_2}{4 \text{ mol HNCO}} \times \frac{6.022 \times 10^{23} \text{ molecules N}_2}{\text{mol N}_2}$$

$$= 9.19 \times 10^{24} \text{ molecules N}_2$$

9.27 $$2.05 \text{ g HNCO} \frac{\text{mol HNCO}}{43.025 \text{ g HNCO}} \times \frac{4 \text{ mol CO}_2}{4 \text{ mol HNCO}} = 0.0476 \text{ mol CO}_2$$

9.29 A recipe calls for two eggs, but you have only one. You will have to use half of the amounts of all the other ingredients listed in the recipe. You will yield only half of the amount indicated in the recipe. Another example is trying to make a sandwich with only one piece of bread.

9.31 $CaO + 3C \rightarrow CaC_2 + CO$

a) Find the limiting reactant by doing two calculations of mass for CaC_2.

$$50.0 \text{ g CaO} \times \frac{1 \text{ mol CaO}}{56.077 \text{ g CaO}} \times \frac{1 \text{ mol CaC}_2}{1 \text{ mol CaO}} \times \frac{64.100 \text{ g CaC}_2}{1 \text{ mol CaC}_2} = 57.153 \text{ g CaC}_2$$

$$34.0 \text{ g C} \times \frac{1 \text{ mol C}}{12.011 \text{ g C}} \times \frac{1 \text{ mol CaC}_2}{3 \text{ mol C}} \times \frac{64.100 \text{ g CaC}_2}{1 \text{ mol CaC}_2} = 60.483 \text{ g CaC}_2$$

Because the calculation beginning with CaO results in a smaller amount of the product CaC_2, CaO is the limiting reactant.

b) 57.2 g CaC_2 is formed.

c) The total mass at the end of the reaction is the same total mass at the beginning of the reaction: 50.0 g + 34.0 g = 84.0 g.

d) At the end of the reaction, the limiting reactant is used up; CaO = 0.0 g.
We have calculated the mass of CaC_2; 57.2 g. We must now calculate the mass of CO formed to determine the mass of C that is left unused.

$$50.0 \text{ g CaO} \times \frac{1 \text{ mol CaO}}{56.077 \text{ g CaO}} \times \frac{1 \text{ mol CO}}{1 \text{ mol CaO}} \times \frac{28.010 \text{ g CO}}{1 \text{ mol CO}} = 24.97 \approx 25.0 \text{ g CO}$$

The mass of the remaining reactant is found by subtracting the known masses at the end of the reaction from the total mass at the beginning.

84.0 – 57.2 – 25.0 = 1.8 g of C are not used.

9.33 See 9.32.

$$0.85 \text{ g NH}_3 \times \frac{\text{mol NH}_3}{17.031 \text{ g NH}_3} \times \frac{4 \text{ mol NO}}{4 \text{ mol NH}_3} \times \frac{30.006 \text{ g NO}}{1 \text{ mol NO}} = 1.49 \approx 1.5 \text{ g NO}$$

$$1.28 \text{ g O}_2 \times \frac{\text{mol O}_2}{31.998 \text{ g O}_2} \times \frac{4 \text{ mol NO}}{5 \text{ mol O}_2} \times \frac{30.006 \text{ g NO}}{1 \text{ mol NO}} = 0.960 \text{ g NO}$$

The maximum amount of NO that can be produced is 0.960 g.

9.35 $2C_2H_6 + 7O_2 \rightarrow 4CO_2 + 6H_2O$

a) We must determine the limiting reagent to determine the amount of product that will be obtained.

$$22.0 \text{ g } C_2H_6 \times \frac{\text{mol } C_2H_6}{30.070 \text{ g } C_2H_6} \times \frac{4 \text{ mol } CO_2}{2 \text{ mol } C_2H_6} \times \frac{44.009 \text{ g } CO_2}{\text{mol } CO_2} = 64.4 \text{ g } CO_2$$

$$16.0 \text{ g } O_2 \times \frac{\text{mol } O_2}{31.998 \text{ g } O_2} \times \frac{4 \text{ mol } CO_2}{7 \text{ mol } O_2} \times \frac{44.009 \text{ g } CO_2}{\text{mol } CO_2} = 12.6 \text{ g } CO_2$$

Here, O_2 is the limiting reactant. The maximum amount of CO_2 that can be produced is 12.6 g.

b) We found that 22.0 g C_2H_6 is 0.732 moles C_2H_6, and 16.0 g O_2 is 0.500 moles O_2. The equation shows that 2 moles C_2H_6 are needed for every 7 moles O_2. This makes O_2 the limiting reactant. We will use it to calculate grams of H_2O produced.

$$0.500 \text{ mol } O_2 \times \frac{6 \text{ mol } H_2O}{7 \text{ mol } O_2} \times \frac{18.015 \text{ g } H_2O}{1 \text{ mol } H_2O} = 7.72 \text{ g } H_2O \text{ produced}$$

c) O_2 was the limiting reactant in both (a) and (b)

9.37 $$\text{Percent yield} = \frac{\text{actual yield}}{\text{theorectical yield}} \times 100$$

The theorectical yield can be obtained once we have the balanced equation:
$Pb(NO_3)_2 + Na_2CrO_4 \rightarrow PbCrO_4 + 2NaNO_3$

$$1.85 \text{ g } Pb(NO_3)_2 \times \frac{\text{mol } Pb(NO_3)_2}{331.2 \text{ g } Pb(NO_3)_2} \times \frac{1 \text{ mol } PbCrO_4}{1 \text{ mol } Pb(NO_3)_2} \times \frac{323.2 \text{ g } PbCrO_4}{\text{mol } PbCrO_4}$$

$$= 1.81 \text{ g } PbCrO_4$$

$$\% \text{ yield} = \frac{1.62 \text{ g } PbCrO_4}{1.81 \text{ g } PbCrO_4} \times 100 = 89.5\%$$

9.39 $2H_2O_2 \rightarrow 2H_2O + O_2$

$$25 \text{ g } H_2O_2 \times \frac{\text{mol } H_2O_2}{34.014 \text{ g } H_2O_2} \times \frac{1 \text{ mol } O_2}{2 \text{ mol } H_2O_2} = 0.37 \text{ mol } O_2$$

9.41 $$0.25 \text{ g Ca} \times \frac{\text{mol Ca}}{40.078 \text{ g Ca}} \times \frac{1 \text{ mol } H_2}{1 \text{ mol Ca}} \times \frac{2.016 \text{ g } H_2}{\text{mol } H_2} \times \frac{\text{mol } H_2}{2.016 \text{ g } H_2} = 6.2 \times 10^{-3} \text{ mol } H_2$$

9.43 This is a limiting reagent problem.

$2HClO_3 + H_2C_2O_4 \rightarrow 2ClO_2 + 2CO_2 + 2H_2O$

$$23.5 \text{ g } H_2C_2O_4 \times \frac{\text{mol } H_2C_2O_4}{90.043 \text{ } H_2C_2O_4} \times \frac{2 \text{ mol } ClO_2}{1 \text{ mol } H_2C_2O_4} \times \frac{67.451 \text{ g } ClO_2}{\text{mol } ClO_2} = 35.2 \text{ g } ClO_2$$

$$15.8 \text{ g } HClO_3 \times \frac{\text{mol } HClO_3}{84.458 \text{ } HClO_3} \times \frac{2 \text{ mol } ClO_2}{2 \text{ mol } HClO_3} \times \frac{67.451 \text{ g } ClO_2}{\text{mol } ClO_2} = 12.6 \text{ g } ClO_2$$

$HClO_3$ is the limiting reactant. The maximum amount of ClO_2 that can be produced is 12.6 g.

9.45 $C_5H_{12} + 8O_2 \rightarrow 5CO_2 + 6H_2O$

a) There will be 0.220 mol CO_2 produced.

$$0.176 \text{ mol } C_5H_{12} \times \frac{5 \text{ mol } CO_2}{1 \text{ mol } C_5H_{12}} = 0.880 \text{ mol } CO_2$$

$$0.352 \text{ mol } O_2 \times \frac{5 \text{ mol } CO_2}{8 \text{ mol } O_2} = 0.220 \text{ mol } CO_2$$

b) O_2 is the limiting reagent. We can calculate the excess moles of C_5H_{12} by determining how much reacted.

$$0.352 \text{ mol } O_2 \times \frac{1 \text{ mol } C_5H_{12}}{8 \text{ mol } O_2} = 0.0440 \text{ mol } C_5H_{12} \text{ reacted}$$

$$0.176 \text{ mol} - 0.0440 \text{ mol} = 0.132 \text{ mol } C_5H_{12} \text{ excess}$$

9.47 $Sr(NO_3)_2 + 2NaOH \rightarrow Sr(OH)_2 + 2NaNO_3$

$$25.0 \text{ g } Sr(NO_3)_2 \times \frac{\text{mol } Sr(NO_3)_2}{211.628 \text{ g } Sr(NO_3)_2} \times \frac{2 \text{ mol NaOH}}{1 \text{ mol } Sr(NO_3)_2} \times \frac{39.997 \text{ g NaOH}}{\text{mol NaOH}} = 9.45 \text{ g NaOH}$$

9.49 $18CO_2 + C_6H_{12}O_6 + 24Fe(OH)_3 \rightarrow 24FeCO_3 + 42H_2O$

$$0.100 \text{ g } C_6H_{12}O_6 \times \frac{\text{mol } C_6H_{12}O_6}{180.156 \text{ g } C_6H_{12}O_6} \times \frac{24 \text{ mol } FeCO_3}{1 \text{ mol } C_6H_{12}O_6} \times \frac{115.855 \text{ g } FeCO_3}{\text{mol } FeCO_3}$$

$$= 1.54 \text{ g } FeCO_3 \text{ produced}$$

9.51 $2NO_{(g)} + O_{2(g)} \rightarrow 2NO_{2(g)}$

$$10.0 \text{ g } NH_4 \times \frac{\text{mol NO}}{30.006 \text{ g NO}} \times \frac{2 \text{ mol } NO_2}{2 \text{ mol NO}} \times \frac{46.005 \text{ g } NO_2}{\text{mol } NO_2} = 15.3 \text{ g } NO_2$$

$$11.0 \text{ g } O_2 \times \frac{\text{mol } O_2}{31.998 \text{ g } O_2} \times \frac{2 \text{ mol } NO_2}{1 \text{ mol } O_2} \times \frac{46.005 \text{ g } NO_2}{\text{mol } NO_2} = 31.6 \text{ g } NO_2$$

The maximum amount of NO_2 produced is 15.3 g.

CHAPTER 10

10.1 One can use the words wavy, circular, or curved to describe objects that are not straight.

10.3 The y-intercept can be obtained by plugging in zero for x.

$$x - 14y = -2$$
$$0 - 14y = -2$$
$$-14y = -2$$
$$y = \frac{-2}{-14} = \frac{1}{7}$$

The x-intercept is $x - 0 = -2$ or $x = -2$.

10.5 The form of the equation of a line is $y = mx + b$, where m is the slope and b is the y-intercept. To find the slope, $m = \frac{K_2 - K_1}{C_2 - C_1}$.

$$m = \frac{573K - 273K}{300°C - 0°C} = \frac{300K}{300°C} = 1 \frac{K}{°C}$$

To find the y-intercept, we can use one of the points. Let's use the point (0, 273) where x is the °C value and y is the K value.

$x = 0$ $\qquad$ $y = 273$

$273 = 1(0) + b$

$b = 273$

The equation, therefore, is $y = 1x + b$ or $K = 1(°C) + 273$.

10.7 If we plug in the units given in the problem into the equation, we get:

$\pi = \kappa C$

$$\text{atm} = \kappa\left(\frac{\text{mol}}{\text{L}}\right)$$

$$\kappa = \frac{\text{atm} \cdot \text{L}}{\text{mol}}$$

10.9 a) The independent variable is moles of benzene. The dependent variable is pressure.

b)

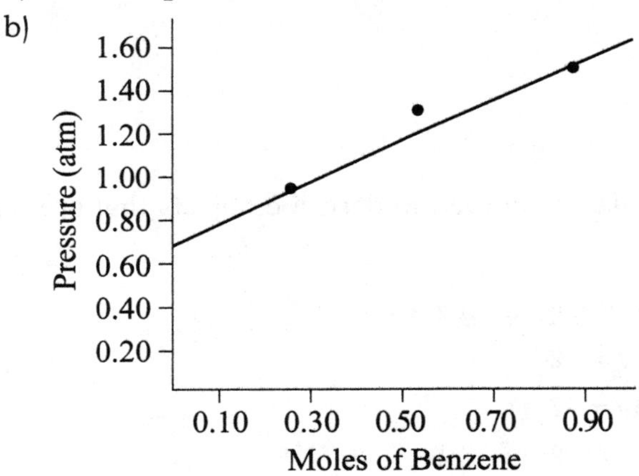

When graphing, the independent variable is on the *x*-axis, while the dependent variable is on the *y*-axis.

c) The slope is $m = \dfrac{\Delta y}{\Delta x} 0.846 = 0.85$

d) The units for the slope are atm · mol^{-1}.

e) The equation of the line generated from this data is:

$$p = (0.85 \frac{\text{atom}}{\text{mol}})(n_{\text{benzene}}) + 0.72 \text{ atm}$$

10.11 See Table 10-2.

a) 1 atm = 760 mmHg

$$0.89 \text{ atm} \times \frac{760 \text{ mmHg}}{1 \text{ atm}} = 680 \text{ mmHg}$$

b) $2.43 \text{ atm} \times \dfrac{760 \text{ mmHg}}{1 \text{ atm}} \times \dfrac{1 \text{ cmHg}}{10 \text{ mmHg}} = 185 \text{ cmHg}$

c) $1.9 \times 10^5 \text{ pascals} \times \dfrac{1 \text{ atm}}{101{,}325 \text{ pascals}} = 1.9 \text{ atm}$

10.13 At constant temperature and moles of gas, the pressure of a fixed amount of gas is inversely proportional to its volume; the larger the volume, the lower the pressure.

$$P_1V_1 = P_2V_2$$

See page 350 for other equations.

10.15 See problem 10.14.

The pressure increased from 0.98 atm to 1.20 atm. We expect the volume to decrease, or be less than 2.00 L.

$P_1 = 0.98$ atm $\qquad V_1 = 2.00$ L

$P_2 = 1.20$ atm $\qquad V_2 = ?$

$(0.98 \text{ atm})(2.00 \text{ L}) = (1.20 \text{ atm})(V_2)$

$$V_2 = \frac{(0.98 \text{ atm})(2.00 \text{ L})}{(1.20 \text{ atm})} = 1.6 \text{ L}$$

This answer makes sense, V_2 is less than 2.00 L.

10.17 To get the same amount of gas at higher pressure, the volume must be decreased. The gas should be put into a smaller container.

10.19 $P_1 = ?$ $\qquad P_2 = 790$ mmHg

$V_1 = 5.0$ L $\qquad V_2 = 6.5$ L

$P_1V_1 = P_2V_2$

$(P_1)(5.0 \text{ L}) = (790 \text{ mmHg})(6.5 \text{ L})$

$$P_1 = \frac{(790 \text{ mmHg})(6.5 \text{ L})}{(5.0 \text{ L})}$$

$P_1 = 1.0 \times 10^3$ mmHg

The answer we get, 1027 mmHg, must be expressed with only two significant figures. This is why it is written in scientific notation.

10.21 $P_1 = 0.88$ atm $\qquad P_2 = ?$

$V_1 = 0.50$ L $\qquad V_2 = 385$ mL

$P_1V_1 = P_2V_2$

$$(0.88 \text{ atm})(0.50 \text{ L}) = (P_2)(385 \text{ mL} \times \frac{1 \text{ L}}{1000 \text{ mL}})$$

$$P_2 = \frac{(0.08 \text{ atm})(0.50 \text{ L})}{(0.385 \text{ L})} = 1.1 \text{ atm}$$

10.23 If we are to have a negative temperature (e.g., –2K) as one of the temperatures in a problem, we would end up calculating a negative volume. This scenario is impossible because you cannot have negative volume.

10.25 See page 354. **REMEMBER:** K = °C + 273.15

a) 10.5°C + 273.15 = 283.65 K = 283.6 K

b) –10.5°C + 273.15 = 262.65 K = 262.6 K

c) 10.5 K =°C + 273.15

°C = 10.5 K – 273.15 = –262.65°C = –262.6°C

d) –182.5°C + 273.15 = 90.65 K = 90.6 K

e)°C = 0.25 K – 273.15 = –272.90°C

10.27 Since there is a fixed amount, one mole, at constant pressure, we use the Charles's Law equation:

$$\frac{V_1}{V_2} = \frac{T_1}{T_2}$$

Remember: T must be in Kelvin (K) and not in Celsius (C).

V_1	T_1	V_2	T_2
220.0 L	–23.5°C = 249.6K	***271.6 L***	35.0°C = 308.1 K
4.2 L	254.53 K	5.0 L	303.18 K
3.50 L	***274.3 K***	7.32 L	300.5°C = 573.6 K
25.0 L	285	0.500 L	***5.7 K***
25.30 L	50.7°C = 323.8 K	***20.2 L***	–15.0°C = 258.1 K
0.0376 L	28.35 K	0.275 L	207.64 K

10.29 The problem gives us 15.0 g Ar, 1.50 L, and 1.05 atm. To get the temperature needed for these conditions, we will use the Ideal Gas Law (PV = nRT). Before we plug our known values into this equation, we must convert the amount of Ar given (grams) into moles.

$$15.0 \text{ g Ar} \times \frac{\text{mol Ar}}{39.948 \text{ g Ar}} = 0.375 \text{ mol}$$

PV = nRT

$$T = \frac{PV}{nR} = \frac{(1.05 \text{ atm})(1.50 \text{ L})}{(0.375 \text{ mol})(0.082057 \frac{\text{L} \cdot \text{atm}}{\text{mol} \cdot \text{atm}})}$$

T = 51.2 K

10.31 The temperature and amount are constant. We need to use Boyle's Law ($P_1V_1 = P_2V_2$).

P_1 = 1.217 atm $\quad$ P_2 = ?

V_1 = 2.50 L $\quad$ V_2 = 4.00 L

$$P_2 = \frac{P_1V_1}{V_2} = \frac{(1.217 \text{ atm})(2.50 \text{ L})}{(4.00 \text{ L})} = 0.761 \text{ atm}$$

10.33 When pressure is doubled, the volume must be halved since P and T are inversely proportional.

10.35 a) Pressure and moles are held constant. We will use Charles's Law ($\frac{V_1}{V_2} = \frac{T_1}{T_2}$).

$$\frac{0.250\ \text{L}}{5.00\ \text{L}} = \frac{273\ \text{K}}{T_2}$$

$T = 5460$ K

b) T and n are constant. We will use Boyle's Law ($P_1V_1 = P_2V_2$),

$(1.0\ \text{atm})(0.250\ \text{L}) = P_2(5.00\ \text{L})$

$P_2 = 0.050$ atm

c) There are three variables: P,V, and T. We will need to use the Combined Gas Law:

$$\frac{P_1V_1}{T_1} = \frac{P_2V_2}{T_2}$$

$$\frac{(1.00\ \text{atm})(0.250\ \text{L})}{(273\ \text{K})} = \frac{(0.050\ \text{atm})(5.00\ \text{L})}{T_2}.$$

$T_2 = 273$ K

d) See part b of this problem. This problem involves Boyle's Law ($P_1V_1 = P_2V_2$).

$P_1(0.250\ \text{L}) = (2.00\ \text{atm})(5.00\ \text{L})$

$P_1 = 40.0$ atm

e) Since n is changing, we need to use the Combined Gas Law($\frac{P_1V_1}{n_1T_1} = \frac{P_2V_2}{n_2T_2}$).

We see that $P_1 = P_2$, and that $T_1 = T_2$. Therefore, the equation we need to use is:

$\frac{V_1}{n_1} = \frac{V_2}{n_2}$ because P_1, P_2, T_1, and T_2 are not necessary in the equation.

$$\frac{0.250\ \text{L}}{2.00\ \text{mol}} = \frac{5.00\ \text{L}}{n_2}$$

$n_2 = 40.0$ mol

f) The number of moles, n, is needed. We must use the Combined Gas Law:

$$\frac{P_1V_1}{n_1T_1} = \frac{P_2V_2}{n_2T_2}.$$

$$\frac{(3.00\ \text{atm})(0.250\ \text{L})}{(2.00\ \text{L})(353\ \text{K})} = \frac{(1.00\ \text{atm})(5.00\ \text{L})}{(2.73\ \text{K})(n_2)}$$

$n_2 = 17.2$ mol

g) The number of moles is held constant. We will use the Combined Gas Law, leaving out n_1 and n_2 (see problem (e) above): $\frac{P_1V_1}{T_1} = \frac{P_2V_2}{T_2}$.

$$\frac{(3.00 \text{ atm})(0.250 \text{ L})}{353 \text{ K}} = \frac{(1.00 \text{ atm})(V_2)}{2.73 \text{ K}}$$

$V_2 = 0.580$ L

10.37 See Figure 10-11. The balanced equation $4Al(s) + 2O_2(g) \rightarrow 2Al_2O_3(s)$ shows a molar ratio of 4 moles Al to 3 moles O_2. If we convert the grams amount of Al to moles, we can get moles O_2. To get volume, we use the Ideal Gas Law (PV = nRT).

$$75.0 \text{ g Al} \times \frac{\text{mol Al}}{26.982 \text{ g Al}} \times \frac{3 \text{ mol } O_2}{4 \text{ mol Al}} = 2.08 \text{ mol } O_2$$

$$V = \frac{nRT}{P} = \frac{(2.08 \text{ mol})(0.082057 \frac{\text{L} \cdot \text{atm}}{\text{mol} \cdot \text{K}})(281 \text{ K})}{3.00 \text{ atm}}$$

V = 16.0 L

10.39 See problem 10.35.

$CH_3OH(l) \rightarrow 2H_2(g) + CO(g)$

$$50.0 \text{ g } CH_3OH \times \frac{\text{mol } CH_3OH}{32.042 \text{ g } CH_3OH} \times \frac{2 \text{ mol H}}{1 \text{ mol } CH_3OH} = 3.12 \text{ mol } H_2$$

$$50.0 \text{ g } CH_3OH \times \frac{\text{mol } CH_3OH}{32.042 \text{ g } CH_3OH} \times \frac{1 \text{ mol CO}}{1 \text{ mol } CH_3OH} = 1.56 \text{ mol CO}$$

a) $$P = \frac{nRT}{V} = \frac{(3.12 \text{ mol } H_2)(0.082057 \frac{\text{L} \cdot \text{atm}}{\text{mol} \cdot \text{K}})(308.35 \text{ K})}{5.00 \text{ L}} = 15.8 \text{ atm}$$

b) $$P = \frac{nRT}{V} = \frac{(1.56 \text{ mol CO})(0.082057 \frac{\text{L} \cdot \text{atm}}{\text{mol} \cdot \text{K}})(308.35 \text{ K})}{5.00 \text{ L}} = 7.90 \text{ atm}$$

c) The total pressure will depend upon the total gas (3.12 mol + 1.56 mol = 4.68).

PV = nRT

$$P = \frac{nRT}{V} = \frac{(4.68 \text{ mol})(0.082057 \frac{\text{L} \cdot \text{atm}}{\text{mol} \cdot \text{K}})(308.35 \text{ K})}{5.00 \text{ L}}$$

P = 23.7 atm

10.41 a) The independent variable is the pressure of Ar because it is the one variable that can be controlled. The dependent variable is solubility because is it the variable that is being determined.

b) y = 0.0026x

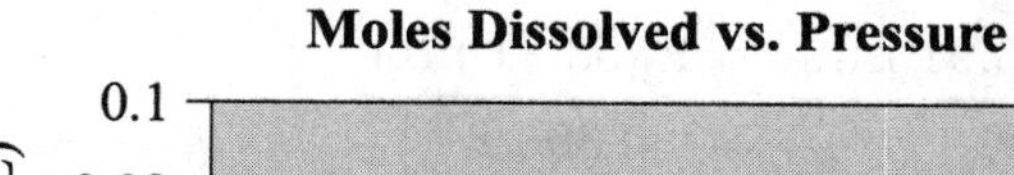

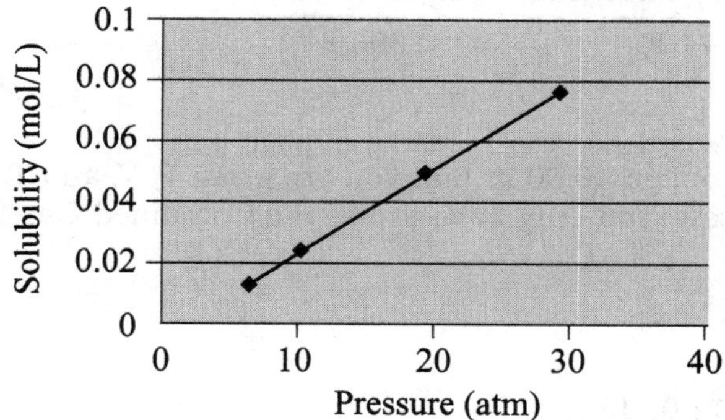

c) Slope: $m = \dfrac{\Delta y}{\Delta x} = \dfrac{0.0512 \text{ mol} - 0.0128 \text{ mol}}{20.0 \text{ atm} - 5.0 \text{ atm}} = 0.00256 \text{ mol atm}^{-1}$ or 2.6×10^{-3} mol.

This can be obtained by using Excel or by using the formula.

d) The units of the slope are mol $L^{-1}atm^{-1}$ or $\dfrac{\text{mol}}{\text{L} \cdot \text{atm}}$.

e) The slope-intercept form of the equation for this line is: $y = 0.00256 \text{ mol atm}^{-1}x + b$.
To get the *y*-intercept we can use one of the data points and solve for b.

$0.0512 \text{ mol} = (0.00256 \text{ mol atm}^{-1})(20.00 \text{ atm}) + b$

$b = 0$

The equation is $s = (0.00256)P$ or $s = 2.6 \times 10^{-3} \left(\dfrac{\text{mol}}{\text{L} \cdot \text{atm}}\right) P(\text{atm})$. Use the variables from the graph.

10.43 The arm span is equal to the height.

10.45 To get the intercepts we must plug in the value of zero.
x-intercept ($y = 0$): $3x - 4(0) = 12$, $x = 4$
y-intercept ($x = 0$): $3(0) - 4y = 12$, $y = -3$

10.47 a) $V = \left(\dfrac{nR}{P}\right)T$

b) $\dfrac{nR}{P}$ is the slope.

c) The *y*-intercept is zero; you can tell from the rearranged equation.

d) The units of slope are $\dfrac{\text{L}}{°\text{C}}$.

10.49 See problem 10.48.

$P_1 = 1.25$ atm $\quad P_2 = ?$

$V_1 = 53.91$ mL $\quad V_2 = 63.81$ mL

$T_1 = 372$ K $\quad T_2 = 398$ K

$$\frac{P_1V_1}{T_1} = \frac{P_2V_2}{T_2} \qquad \frac{(1.25\text{ atm})(53.91\text{ mL})}{371\text{ K}} = \frac{P_2(63.81\text{ mL})}{398\text{ K}}$$

$P_2 = 1.13$ atm

10.51 This problem is different than problem 10.50 in that you are given P, V, and T. There is no need to use the Ideal Gas Law. You only have to use the Combined Gas Law:

$$\frac{P_1V_1}{T_1} = \frac{P_2V_2}{T_2}$$

$$\frac{(1.24\text{ atm})(8.00\text{ L})}{298\text{ K}} = \frac{(0.88\text{ atm})(12.00\text{ L})}{T_2}$$

$T_2 = 317$ K

10.53 The value of $\frac{PV}{nRT}$ will remain 1. If the P value changes, so will the other terms of the equation in order to keep the ratio constant.

10.55 See problem 10.51.

$$\frac{P_1V_1}{T_1} = \frac{P_2V_2}{T_2}$$

$P_1 = 1.06$ atm $\quad P_2 = 0.98$ atm

$V_1 = 24.00$ L $\quad V_2 = ?$

$T_1 = 53.12°C + 273.15 = 326.27$ K $\quad T_2 = 62.94°C + 273.15 = 336.09$ K

$$\frac{(1.06\text{ atm})(24.00\text{ L})}{326.27\text{ K}} = \frac{(0.98\text{ atm})(V_2)}{336.09\text{ K}}$$

$V_2 = 26.74 = 27$

10.57 We must determine the number of moles of CO_2, then we can use the molar ratio from the equation and the molar mass to obtain the grams of HCl.

$$P = 726.0\text{ mmHg} \times \frac{1\text{ atm}}{760\text{ mmHg}} = 0.955\text{ atm}$$

$V = 64$ L

$T = 264$ K

$PV = nRT$

$$(0.955\ \text{atm})(64\ \text{L}) = n(0.082057\ \frac{\text{L} \cdot \text{atm}}{\text{mol} \cdot \text{K}})(264\ \text{K})$$

$n = 2.8$ mol CO_2

$$2.8\ \text{mol}\ CO_2 \times \frac{1\ \text{mol HCl}}{1\ \text{mol}\ CO_2} \times \frac{36.461\ \text{g HCl}}{\text{mol HCl}} = 102.9\ \text{HC1}$$

The answer should have only two significant figures; therefore, we must express the answer in scientific notation: 1.03×10^2 g HCl.

10.59 PV = nRT

$$850.0\ \text{mmHg} \times \frac{(1\ \text{atm})}{760\ \text{mmHg}} \times (386\ \text{L}) = n(0.082057\frac{\text{L} \cdot \text{atm}}{\text{mol} \cdot \text{K}})(395\ \text{K})$$

$n = 13.3$ mol NO_2

$$13.3\ \text{mol}\ NO_2 \times \frac{2\ \text{mol}\ HNO_3}{1\ \text{mol}\ NO_2} \times \frac{63.012\ \text{g}\ HNO_3}{\text{mol}\ HNO_3} = 1680\ \text{g}\ HNO_3$$

10.61 We must use the combined gas law, $\frac{P_1V_1}{T_1} = \frac{P_2V_2}{T_2}$, keeping volume constant at 50.0 L.

$P_1 = 1$ atm — $P_2 = ?$

$T_1 = 273$ K — $T_2 = 85.0 + 273 = 358$ K

$V_1 = 50.0$ L — $V_2 = 50.0$ L

$$\frac{P_1V_1}{T_1} = \frac{P_2V_2}{T_2} \qquad \frac{(1\ \text{atm})(\cancel{50.0\ \text{L}})}{(273\ \text{K})} = \frac{(P_2)\ (\cancel{50.0\ \text{L}})}{(358\ \text{K})}$$

$$P_2 = \frac{(1\ \text{atm})(358\ \text{K})}{(273\ \text{K})} = 1.31\ \text{atm}$$

10.63 a) T and n vary directly with P.

b) V varies inversely with P.

c) R is a constant and doesn't vary.

10.65 We will use the combined gas law equation for this problem: $\frac{P_1V_1}{T_1} = \frac{P_2V_2}{T_2}$.

However, because there is no mention of a change in volume, the values for V_1 and V_2 drop out.

$P_1 = 1$ atm — $P_2 = ?$

$T_1 = 273$ K — $T_2 = 331$ K

$$\frac{P_1}{T_1} = \frac{P_2}{T_2} \qquad P_2 = \frac{P_1 T_2}{T_1} = \frac{(1\ \text{atm})(331.4\ \text{K})}{(273\ \text{K})} = 1.21\ \text{atm or } 923\ \text{mmHg})$$

10.67 $P_1 = 754$ mmHg $\qquad P_2 = 695$ mmHg

$$V_1 = 50.0 \times 10^3\ \text{cm}^3 \times \frac{1\ \text{L}}{10^3\ \text{cm}^3} = 50.0\ \text{L} \qquad V_2 = ?$$

$T_1 = 82.0° + 273 = 355$ K $\qquad T_2 = 35.0 + 273 = 308$ K

$$V_2 = \frac{P_1 V_1 T_2}{T_1 P_2} = \frac{(754\ \text{mmHg})(50.0\ \text{L})(308\ \text{K})}{(355\ \text{K})(695\ \text{mmHg})} = 47.1\ \text{L} = 4.71 \times 10^{-4}\ \text{mL}$$

10.69 a) $$n_{N_2} = \frac{(755\ \text{mmHg} \times \frac{1\ \text{atm}}{760\ \text{mmHg})}(0.250\ \text{L})}{(0.082057\ \text{L atm mol}^{-1}\ \text{K}^{-1})(273\ \text{K})} = 0.011\ \text{mol}$$

b) $$0.011\ \text{mol } N_2 \times \frac{28.014\ \text{g } N_2}{1\ \text{mol } N_2} = 0.31\ \text{g}$$

c) $$D = \frac{0.31\ \text{g}}{0.250\ \text{L}} = 1.2\ \text{g/L} = 1.2 \times 10^{-3}\ \text{g/mL}$$

CHAPTER 11

11.1 a) The campfire loses heat; the water gains heat.

b) The water loses heat; the cool teapot gains heat.

c) The steam loses heat; the turbines gain heat.

11.3 See Table 11-1.

a) The type of energy that is heating the water would be both electromagnetic and thermal. The electromagnetic energy is the light from the sun, and the thermal energy is from the heat of the air around the bag of water.

b) This involves energy (from the microwaves) that causes the water molecules to vibrate faster. This is thermal energy.

c) Flowing water is in motion, which is a form of mechanical energy.

d) The energy produced from the nuclear reaction in the power plant heats the water. This is nuclear energy.

11.5 *Mechanical:* Heat absorbed by water causing it to be pressurized is used to turn the turbines in a power plant.

Electrical: A solar cell absorbs heat and converts it to electricity.

Chemical: A reaction gets cold (decreases in temperature) as it proceeds. It is absorbing heat. An example would be the instant cold packs used to ice injuries on sports teams.

11.7 "A rock lying in the sun warms up."

a) The rock heats up, i.e., increases in heat, because it is absorbing heat from the sun.

b) The heat of the rock is increasing; $q > 0$.

c) The heat is coming from the sun and from the surrounding air and ground.

11.9 When heat is transferred into an object, the process is called *endothermic*.

For problems 11.11–11.17, we need to remember that the solid state is the lowest energy state, followed by the liquid state, and finally the gaseous state, which is the highest in energy. To go from one state to another requires a change in heat. Going from lower to higher energy state will be an endothermic process, whereas going from a higher energy state to a lower one will be an exothermic process.

11.11 When a substance is heated, its atoms and molecules vibrate more.

11.13 The phase change from solid to liquid is endothermic.

11.15 A phase change from gas to liquid is accompanied by a decrease in molecular motion.

11.17 The phase change from liquid to gas is endothermic.

11.19 $15.6 \text{ kJ} \times \dfrac{1000 \text{ J}}{1 \text{ kJ}} = 15{,}600 \text{ J}$

11.21 Water has a higher specific heat capacity. This means it takes more heat to change its temperature than it does to change the temperature of ice. With 100 J of heat, ice will undergo the larger temperature change.

11.23

Substance	Molar Mass (g/mol^{-1})	Specific Heat Capacity ($Jg^{-1}k^{-1}$)	Molar Heat Capacity ($Jmol^{-1}k^{-1}$)
C (diamond)	12.011	2.53	30.4
Si	28.086	0.89	25
Ge	72.61	0.36	26
Sn	118.71	0.22	26
Pb	207.2	0.13	27

11.25 a) Cr: $1.00 \text{ kg Cr} \times \frac{1000 \text{ g Cr}}{1 \text{ kg Cr}} \times \frac{\text{mol Cr}}{51.996 \text{ kg Cr}} \times \frac{21 \text{ kJ}}{\text{mol Cr}} = 404 \text{ kJ}$

Mo: $1.00 \text{ kg Mo} \times \frac{1000 \text{ g Mo}}{1 \text{ kg Mo}} \times \frac{\text{mol Mo}}{95.94 \text{ kg Mo}} \times \frac{28 \text{ kJ}}{\text{mol Mo}} = 292 \text{ kJ}$

Fe: $1.00 \text{ kg Fe} \times \frac{1000 \text{ g Fe}}{1 \text{ kg Fe}} \times \frac{\text{mol Fe}}{55.847 \text{ g Fe}} \times \frac{13.8 \text{ kJ}}{\text{mol Fe}} = 247 \text{ kJ}$

b) from Cr: $404 \text{ kJ} \times \frac{\text{mol } H_2O}{40.66 \text{ kJ}} = 9.94 \text{ mol } H_2O$

$$9.94 \text{ mol } H_2O \times \frac{18.015 \text{ g } H_2O}{\text{mol } H_2O} = 179 \text{ g } H_2O$$

from Mo: $292 \text{ kJ} \times \frac{\text{mol } H_2O}{40.66 \text{ kJ}} = 7.18 \text{ mol } H_2O$

$$7.18 \text{ mol } H_2O \times \frac{18.015 \text{ g } H_2O}{\text{mol } H_2O} = 129 \text{ g } H_2O$$

from Fe: $247 \text{ kJ} \times \frac{\text{mol } H_2O}{40.66 \text{ kJ}} = 6.07 \text{ mol } H_2O$

$$6.07 \text{ mol } H_2O \times \frac{18.015 \text{ g } H_2O}{\text{mol } H_2O} = 109 \text{ g } H_2O$$

11.27 The larger the specific heat capacity, the more heat it takes to change that substance's temperature. To cause a material to undergo a large temperature change when it is placed in a preheated chamber, you should choose a material with a small specific heat capacity. If we assume 100 joules of heat and a 10.0 g sample of material, we can calculate the following change in temperature (ΔT):

H_2O: $100 \text{ J} = (4.18 \frac{\text{J}}{\text{gK}})(10.0 \text{ g})\ \Delta T$; $\Delta T = 2.39 \text{ K}$

Toluene: $100 \text{ J} = (1.69 \frac{\text{J}}{\text{gK}})(10.0 \text{ g})\Delta T$; $\Delta T = 5.92 \text{ K}$

11.29 a) An endothermic reaction is one in which $q > 0$, a positive value. All three substances (Cr_2O_3, MoO_3, and Fe_2O_3) have exothermic heats of reaction.

b) Cr_2O_3: $100 \text{ g } Cr_2O_3 \times \frac{\text{mol } Cr_2O_3}{151.989 \text{ g } Cr_2O_3} \times \frac{-1139.7 \text{ kJ}}{\text{mol}} = -750 \text{ kJ}$

MoO_3: $100 \text{ g } MoO_3 \times \frac{\text{mol } MoO_3}{146.94 \text{ g } MoO_3} \times \frac{-754.5 \text{ kJ}}{\text{mol}} = -513 \text{ kJ}$

Fe_2O_3: $100 \text{ g } Fe_2O_3 \times \frac{\text{mol } Fe_2O_3}{159.691 \text{ g } Fe_2O_3} \times \frac{-822 \text{ kJ}}{\text{mol}} = -515 \text{ kJ}$

b) Although it does not give off the most heat per amount, Fe_2O_3 gives off a good amount of heat (much more than Ag_2O and NiO). It is also more abundant than Cr_2O_3 and MoO_3.

11.31 a) Glucose: $\frac{-2753 \text{ kJ}}{\text{mol}} \times \frac{1 \text{ mol } C_2H_{12}O_6}{180.156 \text{ g } C_2H_{12}O_6} = \frac{-15.28 \text{ kJ}}{\text{g } C_2H_{12}O_6}$

Alanine: $\frac{-1377 \text{ kJ}}{\text{mol}} \times \frac{1 \text{ mol } C_3H_7NO_2}{89.094 \text{ g } C_3H_7NO_2} = \frac{-15.46 \text{ kJ}}{\text{g } C_3H_7NO_2}$

Methyl oleate: $\frac{-10860 \text{ kJ}}{\text{mol}} \times \frac{1 \text{ mol } C_{19}H_{36}O_2}{296.495 \text{ g } C_{19}H_{36}O_2} = \frac{-36.63 \text{ kJ}}{\text{g } C_{19}H_{36}O_2}$

b) Glucose: $100 \text{ g glucose} \times \frac{-15.28 \text{ kJ}}{\text{g glucose}} \times \frac{1 \text{ Cal}}{4.184 \text{ kJ}} = -365 \text{ Cal}$

Alanine: $100 \text{ g alanine} \times \frac{-15.46 \text{ kJ}}{\text{g alanine}} \times \frac{1 \text{ Cal}}{4.184 \text{ kJ}} = -370 \text{ Cal}$

Methyl oleate: $100 \text{ g methyl oleate} \times \frac{-36.63 \text{ kJ}}{\text{g methyl oleate}} \times \frac{1 \text{ Cal}}{4.184 \text{ kJ}} = -875 \text{ Cal}$

11.33 a) The decomposition is exothermic. The heat produced flows to the ice and melts it, forming a liquid layer.

b) Crystallization of the supersaturated solution produces heat—it is exothermic. The heat flows from the pack to your hands as you hold it.

11.35 The butane must be trapped in a small chamber and ignited. As long as it remains in the transport container and not ignited, it will not give off heat.

11.37 a) The heat change for the water can be calculated using the equation: $q = cm\Delta T$.

$q = (4.18 \frac{\text{J}}{\text{gK}})(50 \text{ g})(2.7 \text{ K}) = 564 \text{ J}$

b) The heat change for the water was caused by the reaction. Since heat is given off by the reaction, $q > 0$. Therefore, the heat change of the reaction is -564 J.

c) The heat change we calculated in part (a) is based on 0.010 mol of each reactant. If we double that amount, we will double the heat change, or -1130 J.

d) We expect the temperature change to be twice as much, or 5.4 K.

11.39 See Table 11-4.

$q = cm\Delta T$

$$-5.00 \text{ J} = (.0518 \frac{\text{J}}{\text{gK}})(25.00 \text{ g})\Delta T$$

$\Delta T = -0.39$ K

11.41 $q = cm\Delta T$

$$q = (1.69 \frac{\text{J}}{\text{gK}})(5.0 \text{ g})(25.0°\text{C} - 10.0°\text{C})$$

$q = 127$ J

11.43 *Heat of vaporization* is the amount of heat required to boil a given amount (mol or g) of a liquid.

Heat of fusion is the amount of heat required to melt a given amount (mol or g) of a solid.

11.45 We need to use the heat of fusion when calculating heat transfer in converting a solid to a liquid.

$$10.0 \text{ kg } HC_2H_3O_2 \times \frac{\text{mol } HC_2H_3O_2}{60.052 \text{ g } HC_2H_3O_2} \times \frac{12.09 \text{ kJ}}{\text{mol } HC_2H_3O_2} = 2.01 \text{ kJ}$$

11.47 a) Ni: $q_{Ni} = (0.439 \frac{\text{J}}{\text{g} \cdot \text{K}})(1000 \text{ g})(1455 \text{ K} - 373 \text{ K}) = 475{,}000 \text{ J}$

Ag: $q_{Ag} = (0.225 \frac{\text{J}}{\text{g} \cdot \text{K}})(1000 \text{ g})(961 \text{ K} - 373 \text{ K}) = 132{,}000 \text{ J}$

Au: $q_{Au} = (0.130 \frac{\text{J}}{\text{g} \cdot \text{K}})(1000 \text{ g})(1064 \text{ K} - 373 \text{ K}) = 89{,}800 \text{ J}$

b) from Ni: $475{,}000 \text{ J} \times \frac{1 \text{ kJ}}{1000 \text{ J}} \times \frac{\text{mol } H_2O}{40.66 \text{ kJ}} = 11.7 \text{ mol } H_2O$

$$18.2 \text{ mol } H_2O \frac{18.015 \text{ g } H_2O}{\text{mol } H_2O} = 210 \text{ mol } H_2O$$

from Ag: $132{,}000 \text{ J} \times \frac{1 \text{ kJ}}{1000 \text{ J}} \times \frac{\text{mol } H_2O}{40.66 \text{ kJ}} = 3.25 \text{ mol } H_2O$

$$3.25 \text{ mol } H_2O \frac{18.015 \text{ g } H_2O}{\text{mol } H_2O} = 58.5 \text{ mol } H_2O$$

from Au: $89{,}800 \text{ J} \times \frac{1 \text{ kJ}}{1000 \text{ J}} \times \frac{\text{mol } H_2O}{40.66 \text{ kJ}} = 2.21 \text{ mol } H_2O$

$$2.21 \text{ mol } H_2O \frac{18.015 \text{ g } H_2O}{\text{mol } H_2O} = 39.8 \text{ mol } H_2O$$

11.49 a) An endothermic reaction is one in which $q > 0$, a positive value. The only substance that has an endothermic heat of reaction is Au_2O_3. The other two substances (Ag_2O and NiO) have exothermic heats of reaction.

b) Ag_2O: $100 \text{ g } Ag_2O \times \frac{\text{mol } Ag_2O}{231.74 \text{ g } Ag_2O} \times \frac{-30.6 \text{ kJ}}{\text{mol}} = -13.2 \text{ kJ}$

NiO: $100 \text{ g NiO} \times \frac{\text{mol NiO}}{74.69 \text{ g NiO}} \times \frac{-239.7 \text{ kJ}}{\text{mol}} = -321 \text{ kJ}$

c) Iron oxide is probably the safest and cheapest of all the comparable oxides given. NiO provides more than 16 times the heat of Ag_2O and Au_2O_3.

11.51 a) $q = (4.18 \text{ J K}^{-1} \cancel{\text{g}^{-1}})(10.0 \cancel{\text{g}})(100 \text{ K}) = 4180 \text{ J}$

b) $H_2O(l) \rightarrow H_2O(g)$ not $H_2O(g) \rightarrow H_2O(l)$

$q_{vap} = 40.66 \text{ kJ mol}^{-1}$

$$10.0 \text{ g } H_2O \times \frac{1 \text{ mol } H_2O}{18.015 \text{ g } H_2O} \times \frac{40.66 \text{ kJ}}{\text{mol } H_2O} = 22.6 \text{ kJ}$$

c) $q_{total} = 4.180 \text{ kJ} + 22.6 \text{ kJ} = 26.8 \text{ kJ}$

Here the answer cannot be reported with significant figures beyond the tens place. See significant figure rules.

11.53 Heat capacity of 1 mol $= \frac{136 \text{ J}}{\text{K}}$; specific heat $= \frac{136 \text{ J}}{\text{mol K}} \cdot \frac{1 \text{ mol}}{78 \text{ g}} = 1.74 \text{ J (Kg)}^{-1}$

11.55 $15.0 \cancel{\text{g } C_3H_8} \times \frac{1 \cancel{\text{mol } C_3H_8}}{44.097 \cancel{\text{g } C_3H_8}} \times \frac{214 \text{ kJ}}{1 \cancel{\text{mol } C_3H_8}} = 72.8 \text{ kJ}$

11.57 $q = \text{CM}\Delta\text{T}$

$$\text{m} = \frac{q}{\text{C}\Delta\text{T}} = \frac{(15{,}000 \text{ J})}{(4.18 \frac{\text{J}}{\text{g} \cdot \text{K}})(8.50 \text{ K})} = 422 \text{ g}$$

11.59 a) This is an endothermic reaction.

b) $10.0 \cancel{\text{g } CS_2} \times \frac{1 \cancel{\text{mol } CS_2}}{76.143 \cancel{\text{g } CS_2}} \times \frac{89.7 \text{ kJ}}{1 \cancel{\text{mol } CS_2}} = 11.8 \text{ kJ}$

CHAPTER 12

12.1 Isotopes are atoms of the same element (same number of protons) containing different numbers of neutrons and therefore having different masses.

12.3 Isotopes must have the same charge number, but a different mass number. Choices (b) and (c) are isotopes, $^{16}_{7}X$ and $^{17}_{7}X$. Choices (a) and (d) are the same because the mass number is the same; therefore, they are not isotopes.

12.5 See Example 12.5.

The variables are:

$^{63}_{29}Cu$: $f_1 = 0.6909$ $M_1 = 62.9298$

$^{65}_{29}Cu$: $f_2 = 0.3091$ $M_2 = ?$

$M = 63.546$

$M = f_1 M_1 + f_2 M_2$

$63.546 = (0.6909)(62.9298) + (0.3091)(M_2)$

$M_2 = 64.92$

12.7 a) The number of neutrons is the mass number minus the charge number: $(42 - 18) = 24$.

b) The mass number given with the symbol is expressed as a whole number with no decimals. If we round each of the values given to the nearest whole number we get:

$$41.998 = 42$$

$$42.854 = 43$$

$$41.105 = 41$$

The atom with the mass 41.998 is the one closest in mass to the atom $^{42}_{18}Ar$.

12.9 The masses of the three isotopes 104.26, 103.91, and 104.98, range from 103.91 to 104.98. This means the atomic mass of this element will be within that range. The value in (c), 105.10, does not fall within this range and cannot be this element's atomic mass.

12.11 The number given with the name of the element is the mass number. The charge number can be obtained from the periodic table.

a) $^{230}_{90}Th$

b) $^{224}_{88}Ra$

c) $^{131}_{53}I$

d) $^{72}_{30}Zn$

12.13 The atomic mass calculated is 87.62. This matches the atomic mass of Sr.

12.15 The isotope that contributes most to the atomic mass of the element is the one with the highest relative number (abundance). That would be the X-52 isotope.

12.17 See Table 12-4.

alpha: $^{4}_{2}He$ or $^{4}_{2}\alpha$

beta: $^{0}_{-1}e$ or $^{0}_{-1}\beta$

gamma: $^{0}_{0}\gamma$ or γ

For problems 12.19 to 12.23, we need to make sure that both the total mass number and total charge number on both sides are balanced or equal. See Examples 12.9 through 12.11. The charge number of the missing particles along with the mass number will tell us what the particle is.

12.19 $^{39}_{17}Cl \rightarrow {}^{0}_{-1}\beta + {}^{39}_{18}Ar$

12.21 $^{257}_{102}No \rightarrow {}^{253}_{100}Fm + {}^{4}_{2}He$

12.23 The total mass number should be 15; the total charge number should be 7.

$^{14}_{7}N \rightarrow {}^{14}_{8}O + {}^{0}_{-1}\beta$

12.25 This is an alpha emission. It will involve the isotope giving off or producing an alpha particle along with another particle.

$^{210}_{83}Bi \rightarrow {}^{4}_{2}He + {}^{206}_{81}Tl$

12.27 See problem 12.26.

$^{88}_{35}Br \rightarrow {}^{0}_{-1}\beta + {}^{88}_{36}Kr$

For problems 12.29 and 12.31, we will need to make sure that the total mass number and the total charge number are equal on both sides.

12.29 $^{79}_{36}Kr \rightarrow {}^{0}_{+1}\beta + {}^{79}_{35}Br$

12.31 $^{77}_{36}Kr \rightarrow {}^{0}_{+1}\beta + {}^{77}_{35}Br$

12.33 In Figure 12-13, the left downward diagonal arrow indicates α-decay, while the arrow to the right indicates β-decay. The eight steps are:

1) $^{238}_{92}U \rightarrow {}^{4}_{2}He + {}^{234}_{90}Th$
2) $^{234}_{90}Th \rightarrow {}^{0}_{-1}\beta + {}^{234}_{91}Pa$
3) $^{234}_{91}Pa \rightarrow {}^{0}_{-1}\beta + {}^{234}_{92}U$
4) $^{234}_{92}U \rightarrow {}^{4}_{2}He + {}^{230}_{90}Th$
5) $^{230}_{90}Th \rightarrow {}^{4}_{2}He + {}^{226}_{88}Ra$
6) $^{226}_{88}Ra \rightarrow {}^{4}_{2}He + {}^{222}_{86}Rn$
7) $^{222}_{86}Rn \rightarrow {}^{4}_{2}He + {}^{218}_{84}Po$
8) $^{218}_{84}Po \rightarrow {}^{4}_{2}He + {}^{214}_{82}Pb$

12.35 The term "half-life" refers to the time it takes for half of the originally present sample to undergo decay.

12.37 ^{32}P has a half-life of 14.3 days. That's how long it takes for half of a sample to decay.

12.39 Radiation ionizes matter, or living tissue. This interferes with the function of the tissue either by altering it or destroying it all together. The amount of damage caused depends on the type of radiation, the strength of radiation, and the time of exposure.

12.41 The charge of the number of Cl is 17. The symbol is $^{38}_{17}Cl$.

12.43 See problem 12.42.

^{69}Ga: $f_1 = 0.60108$ $M_1 = 68.925580$

^{71}Ga: $f_2 = 0.39892$ $M_2 = 70.9247005$

$M = (0.60108)(68.925580) + (0.393892)(70.9247005)$

$M = 69.723$

12.45 See problem 12.42.

$f_1 = 0.0582$ $M_1 = 53.9396$

$f_2 = 0.9166$ $M_2 = 55.9349$

$f_3 = 0.0219$ $M_3 = 56.9354$

$f_4 = 0.0033$ $M_4 = 57.9333$

$M = (0.0582)(53.9396) + (0.9166)(55.9349) + (0.0219)(56.9354) + (0.0033)(57.9333)$

$M = 55.847$

12.47 a) $^{4}_{2}He$ is produced. This is an α-emission.

b) $^{0}_{-1}\beta$ is produced. This is a β-emission.

12.49 See the Periodic Table.

a) $^{214}_{82}Pb$

b) $^{103}_{44}Ru$

c) $^{230}_{90}Th$

d) $^{234}_{91}Pa$

12.51 $^{230}_{90}Th \rightarrow {}^{4}_{2}He + {}^{226}_{88}Ra$

12.53 $^{214}_{82}Pb \rightarrow {}^{0}_{-1}\beta + {}^{214}_{83}Bi$

12.55 We need to write the balanced equation.

$^{210}_{83}Bi \rightarrow {}^{206}_{81}Tl + {}^{4}_{2}He$

This is an α-emission.

CHAPTER 13

13.1 In Chapter 2, we learned that the atoms will bond. After all the atoms are bonded, we place the extra valence electrons around the atoms to get them to form an octet. If this leaves any atoms without an octet, we form multiple bonds.

a) Cl_2O $2(7) + 6 = 20e^-$

:Ö:
/ \
:Cl: :Cl:

According to Table 13-1, the molecular structure of this molecule is bent because there are two lone pairs and two single bonds around the central atoms, O, for a total of four electron domains.

b) PF_3 $5 + 3(7) = 26e^-$

:F – P – F:
|
:F:

There are three single bonds and one lone pair around P for a total of four electron domains. The molecular structure is trigonal pyramidal.

c) NOCl $5 + 6 + 7 = 18e^-$

If we draw single bonds between N and Cl and the N and O, we get:

:Cl – N – O:

This molecule will contain a single bond, a double bond, and one lone pair around N. The shape of the molecule is bent.

N
/ ⫽
:Cl O

13.3 a) IF_3 $7 + 3(7) = 28e^-$

I is the central atom.

:F – I – F:
|
:F:

There are five electron domains, three single bonds and two lone pairs. See Figure 13-7 and the example on page 440. The molecular structure is T-shaped.

IF_5 $7 + 5(7) = 42e^-$

I is the central atom.

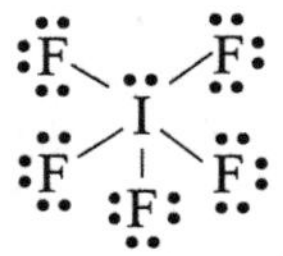

There are six electron domains. The structure of BrF_5 is shown on page 440. The molecular structure is square pyramidal.

b) IF_7 $7 + 7(7) = 56e^-$

I is the central atom.

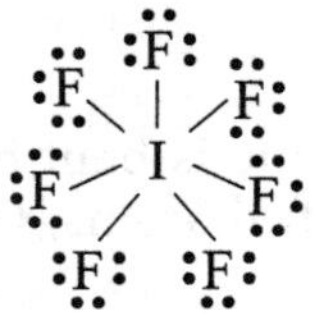

The farthest these domains (7) can be from each other is to have five in one plane with one on top and one on the bottom. This would result in a pentagonal bipyramidal structure.

13.5 SF^- $6 + 7 + 1 = 14e^-$

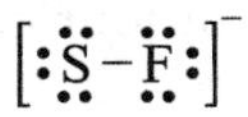

$[:\ddot{\underset{..}{S}}-\ddot{\underset{..}{F}}:]^-$

S is the central atom.

There are four e^- domains around S, one single bond and three lone pairs. The molecular structure is linear.

$SOCl_2$ $6 + 6 + 2(7) = 26e^-$

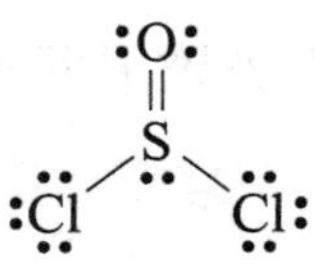

S is the central atom.

This ion is an exception to the octet rule; S is expanded. There are four electron domains around S: one double bond, two single bonds, and one lone pair. The shape is trigonal pyramid.

SF_4 $6 + 4(7) = 34e^-$

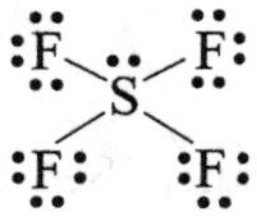

S is the central atom.

There are five electron domains: one lone pair and four single bonds. The molecular structure is see-saw tetrahedral.

H_2S $2(1) + 6 = 8e^-$

S is the central atom.

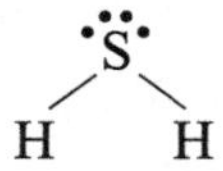

There are four electron domains here: two lone pairs and two single bonds. The molecular structure is bent.

SF_6 $6 + 6(7) = 48e^-$

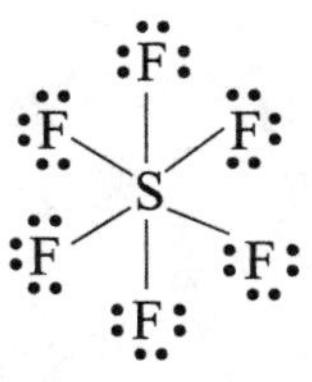

There are six e^- domains around S, all bonding. The molecular structure is octahedral.

SO_4^{2-} $6 + 4(6) + 2 = 32e^-$

S is the central atom.

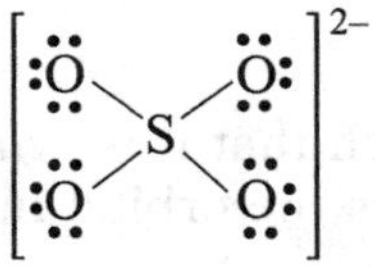

There are four electron domains, all single bonds. The molecular structure is tetrahedral.

13.7 To convert the full structures to compressed structures, we eliminate all C-H bonds and rewrite the carbons with the correct number of hydrogens.

a) 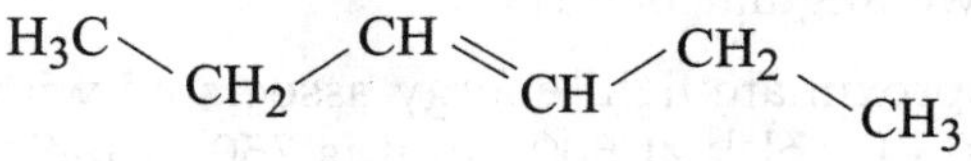 H_3C–CH_2–CH=CH–CH_2–CH_3

b) 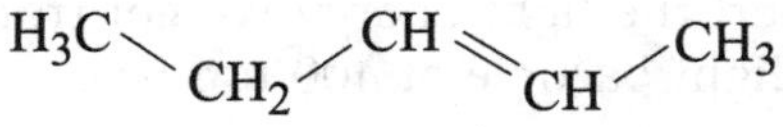 H_3C–CH_2–CH=CH–CH_3

c) 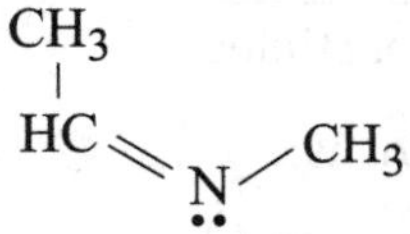 CH_3–HC=N–CH_3

13.9 The full structure simply shows all the C-H bonds.

a)

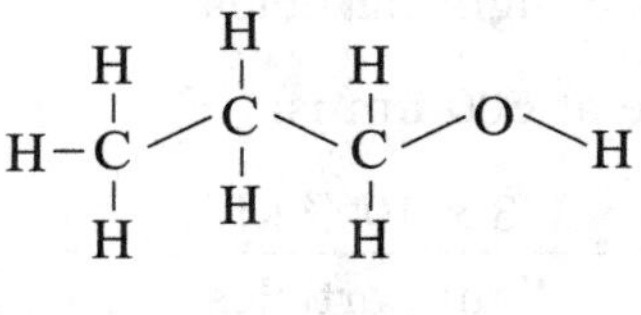

b)

```
             O
 H           ||     H
  \         C       |
H-C  /    \   C /  O  \  H
 H/                |
                   H
```

c)

```
   H         H
   |    O    |
 H-C  /   \  C-H
   |         |
 H-C ------- C-H
   |         |
   H         H
```

d)

```
   H       O
   |    //
 H-C - C
   |     \
   H      O-H
```

e)

```
        O       H
   H    ||      |   H
   |    C       C   |
 H-C  /   \ O /   \ C-H
   |            |   |
   H            H   H
```

13.11 A radial node is one that is a spherical node that is arranged such that it is away from the nucleus—much like layers. An angular node is one that slices the orbital in half. See Table 13-3.

a) 4s There are three radial nodes and no angular nodes. The s-orbital is perfectly spherical.

b) 3p See Figure 13-6. There is one radial node and one angular node.

c) 4d There is one radial node and two angular nodes.

13.13 From Figure 13-12. We see that the approximate light energy associated with one mole of light particles at 400 nm is 300 kJ mol^{-1}; at 500 nm it is 250 kJ mol^{-1}, and at 600 nm it is 200 kJ mol^{-1}. To determine the light energy associated with one single light particle at each wavelength we can convert the light energy we got from Figure 13-12. The energy associated with one single light particle at 400 nm is:

$$\frac{300 \text{ kJ}}{1 \text{ mol light particles}} \times \frac{1 \text{ mol light particles}}{6.022 \times 10^{23} \text{ light particles}} = \frac{5 \times 10^{-22} \text{ kJ}}{\text{light particles}}$$

The energy associated with one single light particle at 500 nm is:

$$\frac{250 \text{ kJ}}{1 \text{ mol light particles}} \times \frac{1 \text{ mol particles}}{6.022 \times 10^{23} \text{ light particles}} = \frac{4 \times 10^{-22} \text{ kJ}}{\text{light particles}}$$

The energy associated with one single light particle at 600 nm is:

$$\frac{200 \text{ kJ}}{1 \text{ mol light particles}} \times \frac{1 \text{ mol light particles}}{6.022 \times 10^{23} \text{ light particles}} = \frac{3 \times 10^{-22} \text{ kJ}}{\text{light particles}}$$

13.15 To determine the proper label for the orbitals refer to Table 13-3.

a) 2 radial nodes, 1 angular node: This node structure is that of a 4p orbital.

b) 3 radial nodes, 0 angular node: The node structure is that of a 4s orbital.

c) 0 radial node, 2 angular nodes: This node structure is that of a 3d orbital.

13.17 Refer to Table 13-3.

a) 3s orbital has 2 radial nodes and no angular nodes.

b) 3d orbital has no radial nodes and 2 angular nodes.

c) 4f orbital has no radial nodes and 3 angular nodes.

13.19 See problem 13-18.

Cs $1s^22s^22p^63s^23p^64s^23d^{10}4p^65s^24d^{10}5p^66s^1$

As $1s^22s^22p^63s^23p^64s^23d^{10}4p^3$

S $1s^22s^22p^63s^23p^4$

I $1s^22s^22p^63s^23p^64s^23d^{10}4p^65s^24d^{10}5p^5$

13.21 Pb $1s^22s^22p^63s^23p^64s^23d^{10}4p^65s^24d^{10}5p^66s^24f^{14}5d^{10}6p^2$

The highest n-value is 6. After the 6s orbital fills, the inner shell d- and f-orbitals fill. Then, the 6p-orbitals begin filling. Pb will lose the two 6p electrons and then it can lose the two 6s electrons. This makes both Pb^{2+} and Pb^{4+} ions possible.

13.23 a) Al $1s^22s^22p^63s^23p^1$

Here the 3p subshell is occupied by one electron.

b) Br $1s^22s^22p^63s^23p^64s^23d^{10}4p^5$

The 4p subshell is occupied by five electrons.

c) N $1s^22s^22p^3$

The 2p subshell is occupied by three electrons.

d) Ti $1s^22s^22p^63s^23p^64s^23d^2$

The 3d subshell is occupied by two electrons.

13.25

O^-	$1s^22s^22p^5$	O	$1s^22s^22p^4$
F^-	$1s^22s^22p^6$ (very stable)	F	$1s^22s^22p^5$
Ne^-	$1s^22s^22p^63s^1$	Ne	$1s^22s^22p^6$ (very stable)
Na^-	$1s^22s^22p^63s^2$	Na	$1s^22s^22p^63s^1$

a) More energy is required when pulling away an electron from a filled subshell that is very stable. We must also consider the resulting atoms configuration. If by pulling away an electron from a monoanion the subshells left are full, the electron affinity will be very low, or zero.

The anion easiest to ionize is Ne^-. The hardest to ionize is F^-. This is the inverse of first ionization energies for the atoms. The atom hardest to ionize is the Ne. This is because Ne has filled subshells. The easiest atom to ionize is Na. When Na loses its valence electron, it is left with filled subshells. This makes Na lose its valence electron easily, or with low ionization energy.

13.27 The diatomic elements are H_2, F_2, Cl_2, Br_2, I_2, O_2, and N_2.

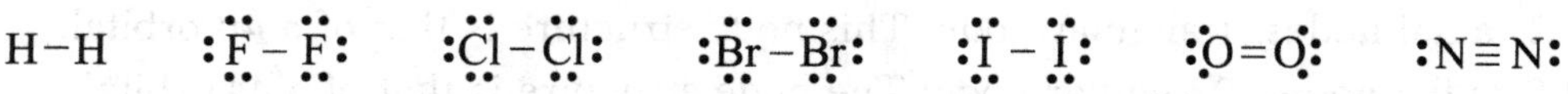

H–H :F–F: :Cl–Cl: :Br–Br: :I–I: :O=O: :N≡N:

13.29 See problem 13.28. See Table 13-1.

SO_4^{2-} $6 + 4(6) = 32e^-$

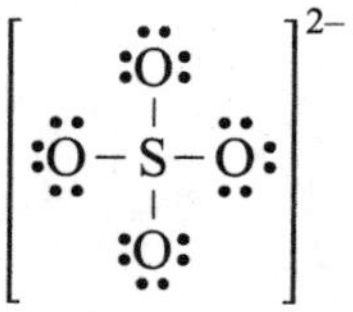

This is tetrahedral. The 3-D structure is also tetrahedral.

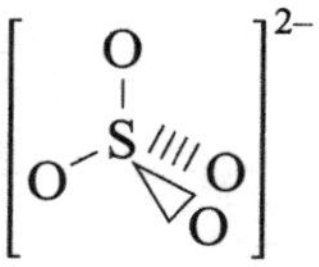

ClO_3^- $7 + 3(6) = 26e^-$

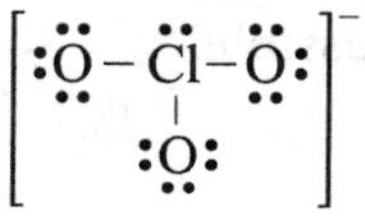

There are three bonding domains and one nonbinding domain. The 3-D structure is trigonal pyramidal.

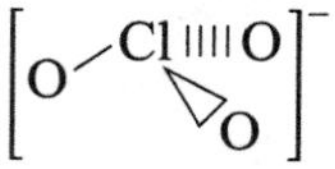

NO_2^- $5 + 2(6) + 1 = 18e^-$

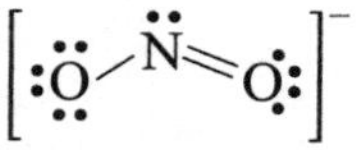

There are two bonding domains and one nonbinding domain. The 3-D structure is bent.

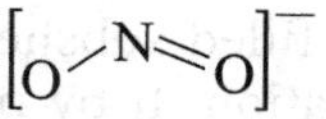

The number of electrons on each is decreasing by 6 – 8 electrons [SO_4^{2-} ($32e^-$), ClO_3^- ($26e^-$), NO_2^- ($18e^-$)] . The number of bonding domains decreased by one each time.

13.31 See Table 13-1.

a)

```
H\     /H
  C=C
H/     \H
```

The carbon atoms in this molecule each have three electron domains, all bonding. We know that the carbon-carbon double bond is planar. The structure is trigonal planar for each C atom.

b)

```
 ① :Ö:      ..②
      \\   .O. ⊖
        C③ ..
        |
 H\ ⊕  C⑤     /H
    N/  |  \C④
 H/ |⑥  H   |  \H
    H       H
```

1 & 2:	The structure here must be linear for each because there is only one bonding electron domain.
3:	The structure is planar because there are three electron domains, all bonding.
4, 5 & 6:	The structure for each is tetrahedral because there are four electron domains, all bonding.

By definition, if we look at any two atoms bonded together, the two atoms will fall in a straight line. We generally consider a three-atom group as the smallest; such as O = C = O. And we consider the 3-D shape around the central atom. For this structure, if we start with the top C as a central atom, we have a structure that is trigonal planar (three bonding domains). Using the second C, we have a tetrahedral shape (four bonding domains). Bottom C on the right is tetrahedral with four bonding domains and if we use the N on the bottom left, we have a tetrahedral structure with four bonding groups.

13.33

a) Si^{4-}	$1s^22s^22p^6\mathbf{3s^23p^6}$	There are eight valence electrons.
b) H^-	$\mathbf{1s^2}$	There are two valence electrons.
c) Br^+	$1s^22s^22p^63s^23p^6\mathbf{4s^2}3d^{10}\mathbf{4p^4}$	There are six valence electrons.
d) B^{3-}	$1s^2\mathbf{2s^22p^4}$	There are six valence electrons.

CHAPTER 14

14.1 When we are dealing with reactions of substances that are in solutions, it is more convenient to use molarity. In reactions, the ratios of substances are in moles. Molarity being $mol \cdot L^{-1}$ would make calculations in reactions more convenient. In the situation of making solutions, it is more convenient to measure the amount of solute in grams or milliliters (for a liquid) rather than have to convert the amount to moles.

14.3 A solution is made up of two components: solute and solvent. The solute is the component that dissolves. The solvent is the component that remains and doesn't change.

14.5 See problem 14.4.

$$mass_{solution} = mass_{HCl} + mass_{H_2O}$$

$$mass_{solution} = 25.0\ g + 500\ g = 525.0\ g$$

14.7 See problem 14.6.

$$mass_{solution} = 22.5\ g + 250.0\ g = 272.5\ g$$

$$272.5\ g\ solution \times \frac{mL\ solution}{1.023\ g\ solution} = 266.4\ mL$$

14.9 In 250.0 mL of 0.760 M HCl, there are $250.0\ mL\ solution \times \frac{0.760\ mol\ HCl}{1000\ mL\ solution}$ $= 0.190\ mol\ HCl$.

14.11 a) To make a 1.23 C_2H_6O solution, 1.23 mol C_2H_6O must be in every 1000 mL. In 500.0 mL solution, there are:

$$500.0\ mL\ solution \times \frac{1.23\ mol\ C_2H_6O}{1000\ mL\ solution} = 0.615\ mol\ C_2H_6O$$

$$0.615\ mol\ C_2H_6O \times \frac{46.069\ mol\ C_2H_6O}{mol\ C_2H_6O} = 28.3\ g\ C_2H_6O$$

b) In 250.0 mL of 8.76×10^{-3} M H_3PO_4 solution, there are:

$$250.0\ mL\ solution \times \frac{8.76 \times 10^{-3}\ mol\ H_3PO_4}{1000\ mL\ solution} = 0.00219\ mol\ H_3PO_4$$

$$0.00219\ mol\ H_3PO_4 \times \frac{97.994\ g\ mol\ H_3PO_4}{mol\ H_3PO_4} = 0.214\ g\ H_3PO_4$$

c) In 3.0 L of 2.5 M $Ca(OCl)_2$ solution, there are:

$$3.0\ L\ solution \times \frac{2.5\ mol\ Ca(OCl)_2}{L\ solution} \times \frac{142.9814\ g\ Ca(OCl)_2}{mol\ Ca(OCl)_2} = 1100\ g$$

14.13 The molarity is determined by dividing the volume of solution, in L, into the amount of solute, in moles.

a) $\frac{25.0\text{ g NaCl}}{2.00\text{ L solution}} \times \frac{\text{mol NaCl}}{58.4427\text{ g NaCl}} = 0.214\text{ M NaCl}$

b) $\frac{0.1052\text{ g } H_2C_2O_4}{1.00\text{ L solution}} \times \frac{\text{mol } H_2C_2O_4}{90.034\text{ g } H_2C_2O_4} = 0.00117\text{ M } H_2C_2O_4$

c) $\frac{0.025\text{ g } PbI_2}{100\text{ mL solution}} \times \frac{1000\text{ mL solution}}{\text{L solution}} \times \frac{\text{mol } PbI_2}{461.000\text{ g } PbI_2} = 0.00054\text{ M } PbI_2$

14.15 See Example 14.8.

a) $0.024\text{ mol NaOH} \times \frac{\text{L solution}}{3.25 \times 10^{-3}\text{ mol NaOH}} = 7.4\text{ L NaOH solution}$

b) $0.198\text{ mol } CH_3COOH \times \frac{\text{L } CH_3COOH}{2.50\text{ mol } CH_3COOH} = 0.0792\text{ L } CH_3COOH\text{ solution}$

c) $1.00 \times 10^{-3}\text{ mol } H_2SO_4 \times \frac{\text{L } H_2SO_4}{0.094\text{ mol } H_2SO_4} = 0.011\text{ L } H_2SO_4\text{ solution}$

14.17 The total mass of the solution is (3.088 g + 25.0 g) = 28.088 g. As we were given d_{soln}, we can solve for v_{soln} using the density formula $d = \frac{m}{v}$, therefore $1.0092\ \frac{g}{mL} = \frac{28.088\text{ g}}{v_{soln}}$ and $v_{soln} = 27.832$ mL. We can get molarity by converting the g KCl to moles and milliliters (mL) solution to liters (L) solution.

$3.088\text{ g KCl} \times \frac{\text{mol KCl}}{74.551\text{ g KCl}} = 0.041421\text{ mol KCl}$

$[KCl] = \frac{0.041421\text{ mol KCl}}{0.027832\text{ L soln}} = 1.488\text{ mol KCl}$

14.19 $BeCl_2(s) \rightarrow Be^{2+}(aq) + 2Cl^-(aq)$

$500.0\text{ mL solution} \times \frac{\text{L solution}}{1000\text{ mL solution}} \times \frac{0.750\text{ mol } BeCl_2}{\text{L solution}} \times \frac{2\text{ mol } Cl^-}{1\text{ mol } BeCl_2}$

$= 0.750\text{ mol } Cl^-$ in 500.0 mL of 0.750 M $BeCl_2$ solution

14.21 We can calculate the moles of acid in each. The reaction is 1 mol acid (HCl or HNO_3) to 1 mol NaOH.

a) $25.00 \text{ mL solution} \times \dfrac{0.982 \text{ mol HCl}}{1000 \text{ mL solution}} \times \dfrac{1 \text{ mol NaOH}}{1 \text{ mol HCl}} \times \dfrac{\text{L solution NaOH}}{0.231 \text{ mol NaOH}}$

$= 0.106 \text{ L NaOH solution} = 106 \text{ L}$

b) $50.00 \text{ L solution} \times \dfrac{0.982 \text{ mol HCl}}{\text{L solution}} \times \dfrac{1 \text{ mol NaOH}}{1 \text{ mol HCl}} \times \dfrac{\text{L NaOH solution}}{0.231 \text{ mol NaOH}}$

$= 213 \text{ L NaOH solution}$

c) $10.0 \text{ L solution} \times \dfrac{2.3 \times 10^{-3} \text{ mol } HNO_3}{\text{L } HNO_3} \times \dfrac{1 \text{ mol NaOH}}{1 \text{ mol } HNO_3} \times \dfrac{\text{L NaOH solution}}{0.231 \text{ mol NaOH}}$

$= 0.0996 \text{ L NaOH solution}$

14.23 See problem 14.21. The $AgNO_3$ solution provides Ag^+, and NaCl provides Cl^-.

$$100.0 \text{ mL } Cl^- \text{ solution} \times \frac{0.00913 \text{ mol } Cl^-}{1000 \text{ mL } Cl^- \text{ solution}} \times \frac{1 \text{ mol } Ag^+}{1 \text{ mol } Cl^-} \times \frac{\text{L } Ag^+ \text{ solution}}{0.0221 \text{ mol } Ag^+}$$

$= 0.0413 \text{ L } AgNO_3 \text{ solution}$

The mass of AgCl(*s*) formed is:

$$100.0 \text{ mL } Cl^- \text{ solution} \times \frac{0.00913 \text{ mol } Cl^-}{1000 \text{ mL } Cl^- \text{ solution}} \times \frac{1 \text{ mol AgCl}}{1 \text{ mol } Cl^-} \times \frac{143.323 \text{ g AgCl}}{\text{mol AgCl}}$$

$= 0.131 \text{ g AgCl}$

14.25 The molarity of OH^- will be mol OH^- per L solution.

$$0.21 \text{ g Ca} \times \frac{\text{mol Ca}}{40.078 \text{ g Ca}} \times \frac{1 \text{ mol } Ca(OH)_2}{1 \text{ mol Ca}} \times \frac{2 \text{ mol } OH^-}{1 \text{ mol } Ca(OH)_2}$$

$= 0.01048 \text{ mol } OH^-$

$$\frac{0.01048 \text{ mol } OH^-}{500.0 \text{ mL solution}} \times \frac{1000 \text{ mL solution}}{\text{L solution}} = 0.021 \text{ M } OH^-$$

14.27 See problem 14.23.

$Ag^+(aq) + Cl^-(aq) \rightarrow AgCl(s)$

$$0.812 \text{ g NaCl} \times \frac{\text{mol NaCl}}{58.443 \text{ g NaCl}} \times \frac{1 \text{ mol } Cl^-}{1 \text{ mol NaCl}} \times \frac{1 \text{ mol } Ag^+}{1 \text{ mol } Cl^-} \times \frac{\text{L solution}}{0.0221 \text{ mol } Ag^+}$$

$= 0.629 \text{ L } AgNO_3 \text{ solution}$

14.29 $Mg^{+} + EDTA^{4-} \rightarrow Mg(EDTA)^{2-}$

a) Moles of EDTA are:

$$23.95 \text{ mL solution} \times \frac{0.1002 \text{ mol EDTA}}{1000 \text{ mL solution}} = 0.00240 \text{ mol EDTA}$$

b) $$0.00240 \text{ mol EDTA} \times \frac{1 \text{ mol } Mg^{2+}}{1 \text{ mol EDTA}} = 0.00240 \text{ mol } Mg^{2+}$$

c) Molarity of Mg^{2+} is:

$$\frac{0.00240 \text{ mol } Mg^{2+}}{50.00 \text{ mL solution}} \times \frac{1000 \text{ mL solution}}{\text{L solution}} = 0.0480 \text{ M } Mg^{2+}$$

14.31 a) Moles HCl added are:

$$100.0 \text{ mL solution} \times \frac{0.193 \text{ mol HCl}}{1000 \text{ mL solution}} = 0.0193 \text{ mol HCl}$$

Moles NH_3 added are:

$$100.0 \text{ mL solution} \times \frac{0.0865 \text{ mol } NH_3}{1000 \text{ mL solution}} = 0.00865 \text{ mol } NH_3$$

b) The NH_3 is the limiting reactant; HCl is excess.

c) There is a one-to-one molar ratio of HCl to NH_3. The amount of HCl reacted is 0.00865 mol. The amount remaining is (0.0193 mol - 0.00865 mol) = 0.0107 mol HCl. The amount of NH_4^+ formed is:

$$0.00865 \text{ mol } NH_3 \times \frac{1 \text{ mol } NH_4^+}{1 \text{ mol } NH_3} = 0.00865 \text{ mol } NH_4^+$$

d) The molarity of the $[NH_4^+]$ is:

$$\frac{0.00865 \text{ mol } NH_4^+}{200.0 \text{ mL solution}} \times \frac{1000 \text{ mL solution}}{\text{L solution}} = 0.0433 \text{ M } NH_4^+$$

The molarity of the excess HCl is:

$$\frac{0.0107 \text{ mol HCl}}{200.0 \text{ mL solution}} \times \frac{1000 \text{ mL solution}}{\text{L solution}} = 0.0535 \text{ M HCl}$$

14.33 In a galvanic cell, oxidation occurs at the anode while reduction occurs at the cathode.

14.35 We must balance the charge on both sides of the equation by adding electrons accordingly. The reaction in which electrons are added (gained) is reduction. The reaction in which electrons are produced (lost) is oxidation.

a) $Cl_2 + 2e^- \rightarrow 2Cl^-$ reduction

b) $Mn^{3+} + 3e^- \rightarrow Mn$ reduction

c) $2H^+ + 2e^- \rightarrow H_2$ reduction

14.37 Find the least common multiple for the number of electrons in both equations.

a) We should multiply the first equation by 3 and the second one by 2 to end up with $6e^-$ in each.

$3Mg \rightarrow 3Mg^{2+} + 6e^-$

$2Cr^{3+} + 6e^- \rightarrow 2Cr$

$3Mg + 2Cr^{3+} \rightarrow 3Mg^{2+} + 2Cr$

b) Multiply the first equation by 3.

$3Cu^+ + 3e^- \rightarrow 3Cu$

$Al \rightarrow Al^{3+} + 3e^-$

$3Cu^+ + Al \rightarrow 3Cu + Al^{3+}$

14.39 Since the charges on both sides are equal along with number of atoms of each element, this redox equation is balanced as written. Ca is oxidized and Zn^{2+} is reduced.

14.41 This equation is not balanced. The charges are not the same.
$2Al(s) + 3Co^{2+}(aq) \rightarrow 2Al^{3+}(aq) + 3Co(s)$

14.43 Oxidation occurs at the anode; reduction occurs at the cathode. The half reactions are:

$Zn(s) \rightarrow Zn^{2+}(aq) + 2e^-$ anode

$Cu^{2+}(aq) + 2e^- \rightarrow Cu(s)$ cathode

Solid Cu is produced at the cathode. It will gain in mass.

14.45

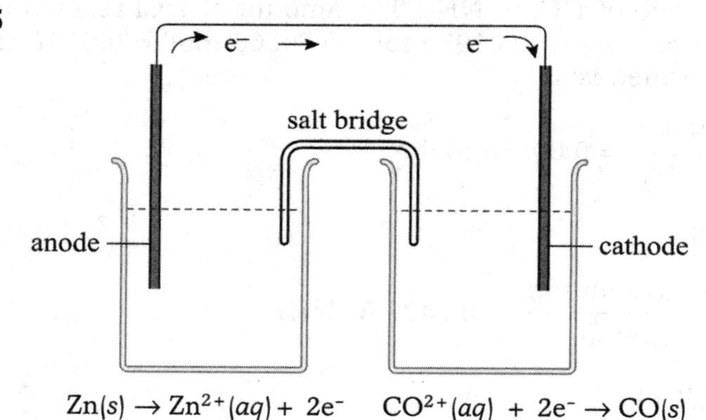

$Zn(s) \rightarrow Zn^{2+}(aq) + 2e^-$ $CO^{2+}(aq) + 2e^- \rightarrow CO(s)$

14.47 a) $K_2Cr_2O_7(s)_l \rightarrow 2K^+(aq) + Cr_2O_7^{2-}(aq)$

One mole $K_2Cr_2O_7$ forms 2 moles K^+ and 1 mole $Cr_2O_7^{2-}$ in solution.

b) $SnBr_2(s) \rightarrow Sn^{2+}(aq) + 2Br^-(aq)$

Two moles $SnBr_2$ form 2 moles Sn^{2+} and 4 moles Br^- in solution.

c) $Co(NO_3)_3(s) \rightarrow Co^{3+}(aq) + 3NO_3^-(aq)$

In solution, 1.2 moles $Co(NO_3)_3$ will form 1.2 moles Co^{3+} and 3.6 moles NO_3^-.

14.49 a) $MnCrO_4(s) \rightarrow Mn^{2+}(aq) + CrO_4^{2-}(aq)$

In solution, 3.5 moles $MnCrO_4$ will form 3.5 moles Mn^{2+} and 3.5 moles CrO_4^{2-}.

b) $Ba(OH)_2 \rightarrow Ba^{2+}(aq) + 2OH^-(aq)$

In solution, 0.8 moles $Ba(OH)_2$ will form 0.8 moles Ba^{2+} and 1.6 moles OH^-.

14.51 $2KI + Pb(NO_3)_2 \rightarrow PbI_2 + 2KNO_3$

According to the equation, one mole of PBI_2 will be produced from 2 moles of KI.

$$125.0 \text{ mL KI} \times \frac{1 \text{ L KI}}{1000 \text{ mL KI}} \times \frac{0.200 \text{ mol KI}}{1 \text{ L KI}} \times \frac{1 \text{ mol } PbI_2}{2 \text{ mol KI}} \times \frac{461.0 \text{ g } PbI_2}{1 \text{ mol } PbI_2} = 5.76 \text{ g } PbI_2$$

14.53 $2\ NaOH + H_2SO_4 \rightarrow 2H_2O + Na_2SO_4$

$$25.00 \text{ mL } H_2SO_4 \times \frac{1 \text{ L } H_2SO_4}{1000 \text{ mL } H_2SO_4} \times \frac{0.982 \text{ mol } H_2SO_4}{1 \text{ L } H_2SO_4} \times \frac{2 \text{ mol NaOH}}{1 \text{ mol } H_2SO_4}$$

$$\times \frac{1 \text{ L NaOH}}{0.426 \text{ mol NaOH}} = 0.115 \text{ L NaOH}$$

14.55 $$36.8 \text{ mL } Mg(OH)_2 \times \frac{1 \text{ L } Mg(OH)_2}{1000 \text{ mL } Mg(OH)_2} \times \frac{0.098 \text{ mol } Mg(OH)_2}{1 \text{ L } Mg(OH)_2} \times \frac{2 \text{ mol } OH^-}{1 \text{ mol } Mg(OH)_2}$$

$$= 0.00722 \text{ mol } OH^-$$

14.57 $$24.0 \text{ mL NaCl} \times \frac{1 \text{ L NaCl}}{1000 \text{ mL NaCl}} \times \frac{1.024 \text{ mol NaCl}}{1 \text{ L NaCl}} \times \frac{1 \text{ mol } Na^+}{1 \text{ mol NaCl}} = 0.0246 \text{ mol } Na^+$$

$$42.8 \text{ mL NaOH} \times \frac{1 \text{ L NaOH}}{1000 \text{ mL NaOH}} \times \frac{0.084 \text{ mol NaOH}}{1 \text{ L NaOH}} \times \frac{1 \text{ mol } Na^+}{1 \text{ mol NaOH}}$$

$$= 0.0036 \text{ mol } Na^+$$

0.0282 mol Na^+ total

14.59 $$2.46 \times 10^{-2} \text{ mol } Ag_2Cr_2O_7 \times \frac{1 \text{ L } Ag_2Cr_2O_7}{0.982 \text{ mol } Ag_2Cr_2O_7} = 0.0251 \text{ L } Ag_2Cr_2O_7$$

14.61 $$34.0 \text{ mL } HNO_3 \times \frac{1 \text{ L } HNO_3}{1000 \text{ mL } HNO_3} \times \frac{0.096 \text{ mol } HNO_3}{1 \text{ L } HNO_3} \times \frac{2 \text{ mol } Pb(NO_3)_2}{4 \text{ mol } HNO_3}$$

$$= 0.00163 \text{ mol } Pb(NO_3)_2$$

14.63 $$74.2 \text{ mL } HNO_3 \times \frac{1 \text{ L } HNO_3}{1000 \text{ mL } HNO_3} \times \frac{0.880 \text{ mol } HNO_3}{1 \text{ L } HNO_3} \times \frac{2 \text{ mol } PbO_2}{4 \text{ mol } HNO_3} \times \frac{331.03 \text{ g } PbO_2}{1 \text{ mol } PbO_2}$$

$$= 10.81 \text{ g } Pb(NO_3)_2$$

14.65

oxidation: Zn(s) → Zn^{2+}(aq)

$$Zn(s) + Cu^{2+}(aq) \rightarrow Zn^{2+}(aq) + Cu(s)$$

reduction: Cu^{2+}(aq) → Cu(s)

This equation is balanced as written.

14.67

oxidation: $Co(s) \rightarrow Co^{2+}(aq)$

$$Co(s) + Pt^{2+}(aq) \rightarrow Co^{2+}(aq) + Pt(s)$$

reduction: $Pt^{2+}(aq) \rightarrow Pt(s)$

This equation is already balanced.

For problem 14.69 the cathode will gain in mass during the course of the reaction because solid metal is formed. Oxidation occurs at the anode while reduction occurs at the cathode.

14.69 $Al(s) \rightarrow Al^{3+}(aq) \rightarrow 3e^-$ oxidation — anode

$Co^{2+}(aq) \rightarrow 2e^- \rightarrow Co(s)$ reduction — cathode gains mass

14.71 $$74.2 \text{ mL } HNO_3 \times \frac{1 \text{ L } HNO_3}{1000 \text{ mL } HNO_3} \times \frac{0.880 \text{ mol } HNO_3}{1 \text{ L } HNO_3} \times \frac{2 \text{ mol } PbO_2}{4 \text{ mol } HNO_3} \times \frac{331.03 \text{ g } PbO_2}{1 \text{ mol } PbO_2}$$

$= 10.81 \text{ g } Pb(NO_3)_2$

14.73 $NaCl(aq) + AgNO_3(aq) \rightarrow AgCl(s) + NaNO_3(aq)$

$$NaCl\text{: } 30 \text{ mL solution} \times \frac{1 \text{ L solution}}{1000 \text{ mL solution}} \times \frac{0.705 \text{ mol NaCl}}{1 \text{ L solution}} = 0.212 \text{ mol NaCl}$$

$$AgNO_3\text{: } 25.0 \text{ mL solution} \times \frac{1 \text{ L solution}}{1000 \text{ mL solution}} \times \frac{0.890 \text{ mol } AgNO_3}{1 \text{ L solution}} = 0.0222 \text{ mol } AgNO_3$$

a) The maximum amount of AgCl that can be produced is 0.0212 mol.

b) Based on the equation, we need a 1:1 molar ratio of reactant. NaCl is the limiting reactant because there are fewer moles NaCl than $AgNO_3$.

14.75 a) $$75.0 \text{ mL solution} \times \frac{1 \text{ L solution}}{1000 \text{ mL solution}} \times \frac{0.0842 \text{ mol } K_2CO_3}{1 \text{ L solution}} = 6.32 \times 10^{-3} \text{ mol } K_2CO_3$$

b) $$125.0 \text{ mL solution} \times \frac{1 \text{ L solution}}{1000 \text{ mL solution}} \times \frac{0.648 \text{ mol LiCl}}{1 \text{ L solution}} = 0.0810 \text{ mol KCl}$$

14.77 $$2.09 \times 10^{-2} \text{ mol } OH^- \times \frac{1 \text{ mol } Ba(OH)_2}{2 \text{ mol } OH^-} \times \frac{1 \text{ L solution}}{0.0464 \text{ mol } Ba(OH)_2} = 0.225 \text{ L}$$

14.79 $$15.4 \text{ mL solution} \times \frac{1 \text{ L solution}}{1000 \text{ mL solution}} \times \frac{1.012 \text{ mol } MgI_2}{1 \text{ L solution}} \times \frac{2 \text{ mol } I^-}{1 \text{ mol } MgI_2} = 0.0312 \text{ mol } I^-$$

14.81 $3Fe(NO_3)_2(aq) + 2Al(s) \rightarrow 3Fe(s) + 2Al(NO_3)_3$

$$20.10 \text{ mL solution} \times \frac{1 \text{ L solution}}{1000 \text{ mL solution}} \times \frac{0.948 \text{ mol } Fe(NO_3)_2}{1 \text{ L solution}} \times \frac{2 \text{ mol } Al(NO_3)_2}{3 \text{ mol } Fe(NO_3)_2}$$

$= 0.0127 \text{ mol } Al(NO_3)_3$

CHAPTER 15

15.1 Limestone is $CaCO_3$. Carbonates react with acids (HA) to produce CO_2 gas according to the equation:

$CaCO_3 + 2HA \rightarrow CaA_2 + H_2O + CO_2$

The solution to which the limestone was added must have been an acid solution.

15.3 According to the Arrhenius definition, an acid increases the concentration of the hydronium ion (H_3O^+) in water by releasing H^+ ions. The Lewis definition of an acid states that an acid is a species that accepts an electron pair.

See Table 15-1 for properties of common acids, which includes corrosive, sour taste, caustic, and slippery to touch.

15.5 a) $HClO_4(aq) + H_2O(l) \rightarrow H_3O^+(aq) + ClO_4^-(aq)$

b) $HBr(aq) + H_2O(l) \rightarrow H_3O^+(aq) + Br^-(aq)$

c) $KOH(aq) \rightarrow K^+(aq) + OH^-(aq)$

d) $H_2SO_4(aq) + H_2O(l) \rightarrow HSO_4^-(aq) + H_3O^+(aq)$

15.7 An indicator is an organic substance that changes color depending on whether it interacts with an acid or a base.

15.9 Gaseous sulfur oxides and nitrogen oxides are produced from the burning of fossil fuels and released into the atmosphere. When it rains, these gases dissolve in the water and form acids that cause the deterioration of our environment.

15.11 $[:\ddot{\underset{..}{Cl}}:]^-$

Cl^- has lone pairs that can be donated to a Lewis acid, thus, acting like a Lewis base.

15.13 $HBr + H_2O \rightarrow Br^- + H_3O^+$

$$:\ddot{\underset{..}{Br}}-H + :\ddot{O}(H)(H) \rightarrow [:\ddot{\underset{..}{Br}}:]^- + [H-\ddot{O}(H)-H]^+$$

15.15 $HNO_3(aq) + H_2O(l) \rightarrow H_3O^+(aq) + NO_3^-(aq)$

The equation shows a 1-to-1 ratio of HNO_3 to H_3O^+. There will be 25 H_3O^+ ions formed in the reaction.

15.17 5.5×10^2 moles of H_2SO_4 will form 5.5×10^2 moles of H_3O^+.

15.19 a) $H_3PO_4(aq) + H_2O(l) \leftrightarrow H_3O^+(aq) + H_2PO_4^-(aq)$

b) $C_6H_5COOH(aq) + H_2O(l) \leftrightarrow H_3O^+(aq) + C_6H_5COO^-(aq)$

c) $H_2CO_3(aq) + H_2O(l) \leftrightarrow H_3O^+(aq) + HCO_3^-(aq)$

d) $HF(aq) + H_2O(l) \leftrightarrow H_3O^+(aq) + F^-(aq)$

15.21 The conjugate base of each will be less an H^+. We can write an equation showing the conjugate base with H^+.

a) $H_2SO_3 \rightarrow HSO_3^- + H^+$
b) $H_2O \rightarrow OH^- + H^+$
c) $H_3O^+ \rightarrow H_2O + H^+$
d) $H_2PO_4^- \rightarrow HPO_4^{2-} + H^+$
e) $HS^- \rightarrow S^{2-} + H^+$
f) $NH_4^+ \rightarrow NH_3 + H^+$

15.23 See Tables 15-5 and 15-6.

a) hydronium ion
b) chloric acid
c) oxalic acid
d) hydrosulfuric acid
e) hydrogen selenate ion
f) phosphoric acid

15.25 See problem 15.21.

a) $HSO_3^- \rightarrow SO_3^{2-} + H^+$
b) $H_2O \rightarrow OH^- + H^+$
c) $HF \rightarrow F^- + H^+$

15.27

```
 [       ..        ]-
 [      :O:        ]
 [   ..  |   ..    ]
 [ H-O - S = O:    ]
 [   ..  |         ]
 [      :O:        ]
 [       ..        ]
```

15.29 See problem 15.28. Make sure the charges on both sides of the equation also balance.

a) $H_2SO_4 + NH_3 \rightarrow HSO_4^- + NH_4^+$
b) $HSO_4^- + CO_3^{2-} \rightarrow SO_4^{2-} + HCO_3^-$
c) $H_3PO_4 + NaOH \rightarrow NaH_2PO_4 + H_2O$
d) $H_2SO_4 + Na_2CO_3 \rightarrow Na_2SO_4 + H_2CO_3$

15.31 Knowing that 0.0562 M $HClO_4$ means $\frac{0.0562 \text{ mol } HCl0_4}{\text{L solution}}$ will help us.

$$850.0 \text{ mL} \times \frac{1\text{L}}{1000 \text{ mL}} \times \frac{0.0562 \text{ mol } HClO_4}{\text{L solution}} = 0.0478 \text{ mol } HClO_4$$

15.33 $Ba(OH)_2 \rightarrow Ba^{2+} + 2OH^-$

$$0.010 \text{ g } Ba(OH)_2 \times \frac{\text{mol } Ba(OH)_2}{171.344 \text{ g } Ba(OH)_2} \times \frac{2 \text{ mol } OH^-}{1 \text{ mol } Ba(OH)_2} = 0.000117 \text{ mol } OH^-$$

$$400.0 \text{ mL} \times \frac{\text{L}}{1000 \text{ mL}} = 0.400 \text{ L}$$

$$M = \frac{\text{mol}}{\text{L}} \qquad \frac{0.000117 \ OH^-}{0.400 \text{ L}} = 0.000292 \text{ M } OH^- = 2.92 \times 10^{-4} \text{ M}$$

15.35 See problem 15.34.

$NaOH + HCl \rightarrow NaCl + H_2O$

$$36.4 \text{ mL solution} \times \frac{1 \text{ L solution}}{1000 \text{ mL solution}} \times \frac{0.491 \text{ mol HCl}}{\text{L solution}} = 0.0179 \text{ mol HCl}$$

$$38.6 \text{ mL solution} \times \frac{1 \text{ L solution}}{1000 \text{ mL solution}} \times \frac{0.442 \text{ mol NaOH}}{\text{L solution}} = 0.0171 \text{ mol NaOH}$$

There is excess HCl; therefore, the final solution will be acidic. 1.8 ml of NaOH are needed for a complete reaction.

15.37 $$125.2 \text{ g HBr} \times \frac{\text{mol HBr}}{80.912 \text{ g HBr}} = 1.547 \text{ mol HBr}$$

$$M = \frac{1.547 \text{ mol HBr}}{15.0 \text{ L solution}} = 0.1032$$

15.39 See problem 15.38.

$$51.3 \text{ mL solution} \times \frac{1 \text{ L solution}}{1000 \text{ mL solution}} \times \frac{0.380 \text{ mol HCl}}{\text{L solution}} = 0.0195 \text{ mol HCl}$$

$$35.8 \text{ mL solution} \times \frac{1 \text{ L solution}}{1000 \text{ mL solution}} \times \frac{0.410 \text{ mol NaOH}}{\text{L solution}} = 0.0147 \text{ mol NaOH}$$

There are (0.0195 mol HCl – 0.0147 mol HCl reacted = 0.0048 mol) mol HCl excess in (51.3 mL + 35.8 mL = 87.1 mL) 87.1 mL total volume.

The solution is acidic with a molarity of $\frac{0.0048 \text{ mol HCl}}{0.0871 \text{ L solution}} = 0.055$ M HCl.

15.41

Number	**Logarithm**
1.78×10^{15}	15.250
2,800,000	6.44
6.37×10^{4}	4.804
0.0352	–1.455
0.000000076	–7.12
8.75×10^{-10}	–9.058

Logarithmic versions of numbers are more uniform, shorter, and easier to use than very large numbers (such as 34,567,123) or very small numbers (such as 0.00000674). The log of 34,000,000 = 7.53, and the log of 0.00000674 = –5.17. Also, if the logarithm value is greater than zero (positive) this tells us the number is greater than one. If the value is less than zero (negative), this tells us the number is less than one.

15.43 See problem 15.42.

Remember that $pH = -\log[H_3O^+]$, $pOH = -\log[OH^-]$ and $pH + pOH = 14$.

a) $pH = 2.5$

$-\log[H_3O^+] = 2.5$

$[H_3O^+] = 10^{-2.5}$

$[H_3O^+] = 0.003\ M = 3 \times 10^{-3}\ M$

b) $pH = 8.234$

$-\log[H_3O^+] = 8.234$

$[H_3O^+] = 10^{-8.234}$

$[H_3O^+] = 5.83 \times 10^{-9}\ M$

c) $[H_3O^+] = 2.7 \times 10^{-9}$

$\log[H_3O^+] = -8.57$

$pH = 8.57$

d) $[H_3O^+] = 0.065$

$\log[H_3O^+] = -1.19$

$pH = 1.19$

e) $[OH^-] = 7.1 \times 10^{-10}$

$\log[OH^-] = -9.15$

$pOH = 9.15$

15.45 We must keep in mind that $pH > 7.0$ is acidic, $pH < 7.0$ is basic, and $pH = 7.0$ is neutral. In terms of concentrations, acidic solutions have $[H_3O^+] > 1.0 \times 10^{-7}$ and $[OH^-] < 1.0 \times 10^{-7}$. Basic solutions have $[H_3O^+] < 1.0 \times 10^{-7}$ M and $[OH^-] > 1.0 \times 10^{-7}$ M.

a) $pH = 10.2$ — This is a basic solution.

b) $pH = 4.33$ — This is an acidic solution.

c) 0.19 M H_2SO_4 — Here, $[H_3O^+] = 0.19$ M, which indicates an acidic solution, $pH = 0.72$.

d) 1.93×10^{-5} M NaOH — The $[OH^-] = 1.93 \times 10^{-5}$ M. Hence, it is a basic solution, $pH = 9.286$.

15.47 $[LiOH] = \dfrac{0.25\ mol}{0.500\ L} = 0.50\ M$

Because LiOH is a strong base:

$[OH^-] = 0.50\ M$

$\log[OH^-] = -0.30$

$pOH = 0.30$

$pH = 14.00 - 0.30 = 13.70$

15.49 a) HCl is a strong acid.

$[HCl] = 0.035\ M$

$[H_3O^+] = 0.035\ M$

$pH = -\log(0.035) = 1.46$

$pOH = 14 - 1.46 = 12.54$

$[OH^-] = 10^{-12.54} = 2.9 \times 10^{-13}\ M$

b) RbOH is a strong base.
$[OH^-] = 3.45 \times 10^{-2}$
$pOH = -\log(3.45 \times 10^{-2}) = 1.462$
$pH = 14.000 - 1.462 = 12.538$
$[H_3O^+] = 10^{-12.538} = 2.90 \times 10^{-13}$

c) H_2SO_4 is a strong acid.
$[H_3O^+]$ is 1.99×10^{-2} M
$pH = -\log(1.99 \times 10^{-2}) = 1.701$
$pOH = 14.000 - 1.701 = 12.299$
$[OH^-] = 10^{-12.299} = 5.03 \times 10^{-13}$ M

15.51 If we consider the acid HA with the base MOH, we have:
$HA + MOH \rightarrow MA + H_2O$
There is a one-to-one molar ratio of acid to base.

$$35.0 \text{ mL solution} \times \frac{1 \text{ L solution}}{1000 \text{ mL solution}} \times \frac{0.15 \text{ mol HA}}{\text{L solution}} = 0.00525 \text{ mol HA}$$

$$25.0 \text{ mL solution} \times \frac{1 \text{ L solution}}{1000 \text{ mL solution}} \times \frac{0.10 \text{ mol MOH}}{\text{L solution}} = 0.0025 \text{ mol MOH}$$

There are (0.00525 mol HA - 0.0025 mol HA reacted = 0.00275 mol) 0.00275 mol HA left in (35.0 mL + 25.0 mL = 60.0 mL) 60 mL total solution.

$$[HA] = [H_3O^+] = \frac{0.00275 \text{ mol}}{0.060 \text{ L}} = 0.0458 \text{ M}$$

$pH = -\log[H_3O^+] = 1.34$

15.53 a) The smallest pH value is that of the solution with the largest hydronium ion concentration in which case the pH would be equal to zero, from problem 15.52a.

b) The largest pH is 11.00, from problem 15.52h.

c) As pH decreases, hydronium ion $[H_3O^+]$ concentration increases.

15.55 a) This question should be answered. The pH can be measured to determine whether the solution is basic or acidic.

b) This question does not need to be answered.

c) This question does not need to be answered.

d) This question should be answered. From the pH measured, one can calculate moles of acid or base present.

e) This question should be answered. It can be measured by using a pH meter or pHydrion paper (*i.e.*, litmus paper).

15.57 $$[HI] = \frac{3.05 \times 10^{-2} \text{ mol HI}}{350.00 \text{ mL solution} \times \frac{1 \text{ L solution}}{1000 \text{ mL solution}}} = 0.0871 \text{ M}$$

HI is a strong acid. Therefore, $[H_3O^+] = 0.0871$ M
$pH = -\log(0.0871) = 1.060$

15.59 The solution contains:

$$35.0 \text{ mL solution} \times \frac{1 \text{ L solution}}{1000 \text{ mL solution}} \times \frac{0.963 \text{ mol HCl}}{\text{L solution}} = 0.0337 \text{ mol HCl}$$

When adding water to the solution we still have 0.0337 mol HCl, but now it is in a volume of 50.0 mL or 0.0500 L.

$$[\text{HCl}] = [\text{H}_3\text{O}^+] = \frac{0.0337 \text{ mol}}{0.0500 \text{ mol}} = 0.674 \text{ M}$$

$\text{pH} = -\log(0.674) = 0.171$

15.61 a) $[\text{H}_3\text{O}^+]$ 9.54×10^{-4} M

When diluted by a factor of 2, $[\text{H}_3\text{O}^+] = \dfrac{9.54 \times 10^{-4} \text{ M}}{2} = 4.77 \times 10^{-4}$

b) $[\text{OH}^-] = 4.2 \times 10^{-5}$ M

After dilution, $[\text{OH}^-] = \dfrac{4.2 \times 10^{-5} \text{ M}}{5} = 8.4 \times 10^{-6}$ M

$$[\text{H}_3\text{O}^+] = \frac{1.0 \times 10^{-14}}{[\text{OH}^-]} = \frac{1.0 \times 10^{-14}}{8.4 \times 10^{-6}} = 1.2 \times 10^{-9} \text{ M}$$

c) pH = 5.44
$[\text{H}_3\text{O}^+] = 10^{-5.44} = 3.6 \times 10^{-6}$ M
After dilution, $[\text{H}_3\text{O}^+] = 1.8 \times 10^{-6}$ M
$\text{pH} = -\log(1.8 \times 10^{-6}) = 5.74$

d) pH = 12.42
$[\text{H}_3\text{O}^+] = 10^{-12.42} = 3.8 \times 10^{-13}$ M
After dilution, $[\text{H}_3\text{O}^+] = 7.6 \times 10^{-14}$ M
$\text{pH} = -\log(7.6 \times 10^{-14}) = 13.12$
$\text{pOH} = 14.00 - 13.12 = 0.88$

15.63 a) pOH = 5.33
$-\log[\text{OH}^-] = 5.33$
$[\text{OH}^-] = 10^{-5.33} = 4.7 \times 10^{-6}$ M

b) $[\text{OH}^-] = 9.2 \times 10^{-3}$ M
$[\text{H}_3\text{O}^+][\text{OH}^-] = 1.0 \times 10^{-14}$

$$[\text{H}_3\text{O}^+] = \frac{1.0 \times 10^{-14}}{9.2 \times 10^{-3}} = 1.1 \times 10^{-12}$$

$\log[\text{H}_3\text{O}^+] = -11.96$
pH = 11.96

c) pH = 4.53
$[\text{H}_3\text{O}^+] = 10^{-4.53}$
$[\text{H}_3\text{O}^+] = 2.95 \times 10^{-5}$

$$[H_3O^+][OH^-] = 1.0 \times 10^{-14}$$

$$[OH^-] = \frac{1.0 \times 10^{-14}}{2.95 \times 10^{-5}}$$

$[OH^-] = 3.39 \times 10^{-10}$
$\log[OH^-] = -9.47$
$pOH = 9.47$

d) $pOH = 6.43$
$pH + pOH = 14$
$pH = 14 - 6.43 = 7.57$

e) $pOH = 1.23$
$pH = 14 - 1.23 = 12.77$
$[H_3O^+] = 10^{-12.77} = 1.7 \times 10^{-13}$ M

15.65 a) $12.48 \text{ g HI} \times \frac{1 \text{ mol HI}}{127.912 \text{ g HI}} = 0.09757 \text{ mol HI}$

b) $0.09757 \text{ mol HI} \times \frac{\text{L}}{0.560 \text{ mol HI}} = 0.174 \text{ mL}$

15.67 a) HI—hydroiodic acid
b) HIO_4—periodic acid
c) HIO_2—iodous acid

15.69 a) HCO_3^- is bicarbonate. Its conjugate acid is H_2CO_3. Its conjugate base is CO_3^{2-}.
b) HCO_3^- $1 + 4 + 3(6) + 1 = 24 \ e^-$

$$\left[\text{H}-\ddot{\underset{..}{\text{O}}}-\underset{\underset{\displaystyle :\ddot{\text{O}}:}{|}}{\text{C}}=\ddot{\underset{..}{\text{O}}} \right]^-$$

c) Looking at the Lewis structures of the reactants,

$$\left[\text{H}-\ddot{\underset{..}{\text{O}}}:\right]^- \quad \text{and} \quad \ddot{\underset{..}{\text{O}}}=\text{C}=\ddot{\underset{..}{\text{O}}}$$

we see that OH^- is providing the lone pair for HCO_3^- to form with CO_2 accepting the lone pair. OH^- is the Lewis base; CO_2 is the Lewis acid.

15.71 a) $C_8H_5O_4^{2-}(aq) + OH^-(aq) \rightarrow H_2O + C_8H_4O_4^-(aq)$

b) The $:\ddot{\underset{..}{\text{O}}}\text{H}^-$ provides a lone pair for H_2O to form; hence, it is the Lewis base.

$C_8H_5O_4^{2-}$ is the Lewis acid.

c) OH^- is the Arrhenius base, by definition, leaving $C_8H_5O_4^{2-}$ as the Arrhenius acid.

d) $$\frac{0.486 \, \cancel{\text{g}} \times \frac{1 \text{ mol } \cancel{C_8H_5O_4K}}{204.222 \text{ g } \cancel{C_8H_5O_4K}} \times \frac{1 \text{ mol NaOH}}{1 \text{ mol } \cancel{C_8H_5O_4K}}}{0.02000 \text{ L}} = 0.119 \text{ M}$$

15.73 $NaOH + HCl \rightarrow H_2O + NaCl$

a) $0.02382 \text{ L HCl} \times \dfrac{0.205 \text{ mol HCl}}{\text{L HCl}} \times \dfrac{1 \text{ mol NaOH}}{1 \text{ mol HCl}} = 0.00488 \text{ mol NaOH}$

b) $0.02382 \text{ L HCl} \times \dfrac{0.103 \text{ mol HCl}}{\text{L HCl}} \times \dfrac{1 \text{ mol NaOH}}{1 \text{ mol HCl}} = 0.00245 \text{ mol NaOH}$

c) $0.02382 \text{ L HCl} \times \dfrac{0.410 \text{ mol HCl}}{\text{L HCl}} \times \dfrac{1 \text{ mol NaOH}}{1 \text{ mol HCl}} = 0.00977 \text{ mol NaOH}$

d) Volume and concentration of the HCl used in the titration are needed to determine moles of NaOH.

CHAPTER 16

16.1 An equilibrium system will have constant concentrations of all reactants and products, whereas the concentrations of reactants and products in a system not at equilibrium will change with time.

16.3 Some macroscopic properties indicating a chemical reaction has occurred are: color change, formation of a gas (bubbles), and formation of an insoluble substance.

16.5 In the equilibrium expression, solution concentrations are expressed in molarity.

16.7 a) $K = \dfrac{[H_3O^+][F^-]}{[HF]}$.

b) $K = [Ca^{2+}][C_2O_4^{2-}]$

c) $K = [Pb^{3+}]^2[CO_3^{2-}]^3$

16.9 $CaC_2O_4(s) \rightleftharpoons Ca^{2+}(aq) + C_2O_4^{2-}(aq)$

$K = [Ca^{2+}][C_2O_4^{2-}] = (3.0 \times 10^{-6})(3.0 \times 10^{-6}) = 9.0 \times 10^{-12}$

16.11 A hydrated metal ion is one that is surrounded by H_2O molecules held in place by the weak attraction between the positively charged metal ion and the lone pairs of electrons on the oxygen in H_2O.

16.13 Drawing the Lewis structure will help us identify the Lewis acid and the Lewis base in $Cu(NH_3)_4^{2+}$.

```
        NH3                    [         NH3          ]2+
        ••                     [          |           ]
H3N : Cu2+ : NH3   →           [ H3N  –  Cu  –  NH3   ]
        ••                     [          |           ]
        NH3                    [         NH3          ]
```

Cu^{2+} is the Lewis acid because it accepts the electron pair. NH_3 is the Lewis base because it donates the electron pair.

16.15

```
 ••
:O – H
 ••
 |
 H
```

There are two lone pairs of electrons on O in H_2O.

16.17 A quadratic equation can be solved using the quadratic formula. Although there are two mathematical solutions to each quadratic equation, these solutions may not be possible. For example, solutions of negative volumes or concentrations are not possible.

16.19 $K = \frac{[HA]}{[A^-]} = 5.0$

The ratio is greater than one. This means the value of the numerator is greater than the volume of the denominator or $[HA] > [A^-]$.

16.21 The units of K will vary depending on the expression. For mathematical reasons beyond the scope of this course, K cannot have units. We divide each concentration in the expression by the standard value of 1 M (or 1 atm for expressions with pressure).

16.23 $Mg^{2+}(aq) + C_3H_5O_3^-(aq) \rightleftharpoons Mg(C_3H_5O_3)^+(aq)$

$$K = \frac{[Mg(C_3H_5O_3)^+]}{[Mg^{2+}][C_3H_5O_3^-]} = \frac{(0.095)}{(0.105)(0.105)} = 8.62$$

16.25 See problem 16.24.

	Mg^{2+}	+	$C_3H_5O_3^-$	$\rightleftharpoons$	$Mg(C_3H_5O_3)^+$
Initial concentration	0.000 M		0.000 M		0.200 M
Change	+ **x**		+ **x**		− **x**
Equilibrium concentration	**x**		**x**		0.200 − **x**

$$8.62 = \frac{0.200 - x}{x^2}$$

$8.62x^2 = 0.200 - x$

$8.62x^2 + x - 0.200 = 0$

$a = 8.62 \qquad b = 1 \qquad c = -0.200$

$$x = \frac{-1 \pm \sqrt{(1) - 4(8.62)(-0.200)}}{2(8.62)} = \frac{-1 \pm 2.81}{17.24}$$

$x = 0.105$, $x = -0.221$ (this value doesn't make sense)

$[Mg^{2+}] = [C_3H_5O_3^-] = 0.105$ M

$[Mg(C_3H_5O_3)^+] = 0.095$ M

16.27 See problems 16.24–16.26.

	Sn^{2+}	+	$C_3H_5O_3^-$	$\rightleftharpoons$	$Sn(C_3H_5O_3)^+$
Initial concentration	0.200 M		0.200 M		0.000 M
Change	− **x**		− **x**		+ **x**
Equilibrium concentration	0.200 − **x**		0.200 − **x**		**x**

$$\frac{x^2}{(0.200 - x)(0.200 - x)} = \frac{x^2}{(0.0400 - 0.400x + x^2)} = 4.92$$

$x^2 = 0.197 - 1.97x + 4.92x^2$

$0 = 0.197 - 2.97x + 4.92x^2$

$a = 4.92 \qquad b = -2.97 \qquad c = 0.197$

$$x = \frac{-b \pm \sqrt{b^2 - 4ac}}{2a}$$

$$x = \frac{2.97 \pm \sqrt{(-2.97)^2 - 4(4.92)(0.197)}}{2(4.92)}$$

$x = 0.528$ (this answer is too big), $x = 0.079$

The value $x = 0.528$ does not make sense because it is larger than the intial concentration.

$[Sn^{2+}] = [C_3H_5O_3^-] = 0.200 - 0.079 = 0.124$ M

$[Sn(C_3H_5O_3)^+] = x = 0.0759$ M

16.29 $PbCO_3(s) + HT^{2-}(aq) \rightleftharpoons PbT^-(aq) + HCO_3^-(aq)$

a) $K = 4.06 \times 10^{-2} = \frac{[PbT^-][HCO_3^-]}{[HT^{2-}]}$

b) $\frac{[PbT^-]}{[HT^{2-}]} = \frac{4.06 \times 10^{-2}}{[HCO_3^-]} = \frac{4.06 \times 10^{-2}}{1.00 \times 10^{-3}} = 40.6$

c) The $\frac{[PbT^-]}{[HT^{2-}]}$ ratio is inversely proportional to $[HCO_3^-]$. As $[HCO_3^-]$ decreases, the ratio will increase.

$$\frac{[PbT^-]}{[HT^{2-}]} = \frac{4.06 \times 10^{-2}}{1.00 \times 10^{-5}} = 4.06 \times 10^3$$

16.31 $ClO^-(aq) + H_2O(l) \rightleftharpoons HClO(aq) + OH^-(aq)$

$K_b = 3.3 \times 10^{-7}$

a) $K = 3.3 \times 10^{-7} = \frac{[HClO][OH^-]}{[ClO^-]}$

b) $\frac{[ClO^-]}{[HClO]} = \frac{[OH^-]}{3.3 \times 10^{-7}} = (3.03 \times 10^6)[OH^-]$

c) pH = 10.00 pOH = 14.00 − 10.00 = 4.00

$[OH^-] = 10^{-4}$

$$\frac{[ClO^-]}{[HClO]} = (3.03 \times 10^6)(10^{-4}) = 3.03 \times 10^2$$

d) $\frac{[ClO^-]}{[HClO]} = (3.03 \times 10^6) = [OH^-]$

From this relationship, we see that the ratio is directly proportional to $[OH^-]$. The higher the $[OH^-]$, the larger the ratio will be, and vice versa.

pH = 3.00 pOH = 14.00 − 3.00 = 11.00

$[OH^-] = 10^{-11}$

$$\frac{[ClO^-]}{[HClO]} = (3.03 \times 10^6) = (10^{-11}) = 3.03 \times 10^{-5}$$

e) An Arrhenius base is a species that increases [OH–] in water. pH measures the acidity of a solution as $pH = -\log[H_3O^+]$. At pH < 7, solutions contain more H_3O^+ ions than OH^- ions; therefore, they are acidic. At pH > 7, solutions contain more OH^- ions than H_3O^+ ions; therefore, they are basic.

pH = 10 means that it is a basic solution.

pH = 3 means that it is an acidic solution.

16.33 a) $CH_3COOH(aq) + H_2O(l) \rightleftharpoons CH_3COO^-(aq) + H_3O^+(aq)$

$$K = 6.5 \times 10^{-5} = \frac{[CH_3COOH^-][H_3O^+]}{[CH_3COOH]}$$

b) $HF(aq) + H_2O(l) \rightleftharpoons H_3O^+(aq) + F^-(aq)$

$$K_a = 6.6 \times 10^{-4} = \frac{[H_3O^+][F^-]}{[HF]}$$

c) $NH_4^+(aq) + H_2O(l) \rightleftharpoons H_3O^+(aq) + NH_3(aq)$

$$K_a = 5.6 \times 10^{-10} = \frac{[H_3O^+][NH_3]}{[NH_4^+]}$$

d) $H_2PO_4^-(aq) + H_2O(l) \rightleftharpoons H_3O^+(aq) + HPO_4^{2-}(aq)$

$$K_a = 6.2 \times 10^{-8} = \frac{[H_3O^+][HPO^2{}_4^-]}{[H_2PO_4^-]}$$

16.35 From Table 16-2:

Acid	K_a	pK_a
HCO_2	1.1×10^{-2}	1.96
H_3PO_4	7.5×10^{-3}	2.12
HF	6.6×10^{-4}	3.18
HCOOH	8.0×10^{-4}	3.10
CH_3COOH	6.5×10^{-5}	4.19
H_2CO_3	4.3×10^{-7}	6.37
$H_2PO_4^-$	6.2×10^{-8}	7.21
HClO	3.0×10^{-8}	7.52
HCN	6.2×10^{-10}	9.21
NH_4^+	5.6×10^{-10}	9.25
HCO_3^-	4.8×10^{-11}	10.32
HPO_4^{2-}	2.2×10^{-13}	12.66

For the general formula of an acid, HA, the pH is related to pK_a as follows.

$HA(aq) + H_2O(l) \rightleftharpoons H_3O^+(aq) + A^-(aq)$

$$K_a = \frac{[H_3O^+][A^-]}{[HA]}$$

Taking the –log of both sides we get:

$$-\log K_a = -\log [H_3O^+] + -\log \frac{[A^-]}{[HA]}$$

$$pK_a = pH - \log \frac{[A^-]}{[HA]}$$

This shows us that the value of concentration of the conjugate base, the concentration of acid, and the concentration of K_a will affect pH.

16.37 See Table 5-2. All these reactions are double-displacement reactions.

a) $2AgNO_3(aq) + Na_2CO_3(aq) \rightarrow 2NaNO_3(aq) + Ag_2CO_3(s)$

b) $CaF_2(aq) + H_2SO_4(aq) \rightarrow 2HF(aq) + CaSO_4(s)$

c) $2KI(aq) + Pb(NO_3)_2(aq) \rightarrow PbI_2(s) + 2KNO_3(aq)$

d) $3CuSO_4(aq) + 2Na_3PO_4(aq) \rightarrow 3Na_2SO_4(aq) + Cu_3(PO_4)_2(s)$

e) $H_2SO_4(aq) + PbCl_2(aq) \rightarrow PbSO_4(s) + 2HCl(aq)$

f) $3H_2SO_4(aq) + Ca_3(PO_4)_2(aq) \rightarrow CaSO_4(s) + 2H_3PO_4(aq)$

g) $K_2SO_4(aq) + Ba(ClO_4)_2(aq) \rightarrow BaSO_4(s) + 2KClO_4(aq)$

16.39 a) $Pb^{2+}(aq) + 2Cl^-(aq) \rightarrow PbCl_2(s)$

b) $Pb^{2+}(aq) + C_2O_4^{2-}(aq) \rightarrow PbC_2O_4(s)$

c) $Mn^{2+}(aq) + 2OH^-(aq) \rightarrow Mn(OH)_2(s)$

d) $Pb^{2+}(aq) + IO_3^-(aq) \rightarrow Pb(IO_3)_2(s)$

16.41 The solubility is the amount of ions in the solution.

a) $HgS(s) \rightleftharpoons Hg^{2+}(aq) + S^{2-}(aq)$

K_{sp} $2.0 \times 10^{-53} = [Hg^{2+}][S^{2-}]$

Only x amount of each product will form. x will be the solubility.

$2.0 \times 10 - 53 = (x)(x)$

$x = \sqrt{2.0 \times 10^{-53}} = 4.47 \times 10^{-27} \approx 4.5 \times 10^{-27}$ M

b) $ZnCO_3(s) \rightleftharpoons Zn^{2+}(aq) + CO_3^{2-}(aq)$

$1.2 \times 10^{-10} = (x)(x)$

$x = \sqrt{1.2 \times 10^{-10}} = 1.10 \times 10^{-5} \approx 1.1 \times 10^{-5}$ M

c) $NiS(s) \rightleftharpoons Ni^{2+}(aq) + S^{2-}(aq)$

$1.1 \times 10^{-21} = (x)(x)$

$x = \sqrt{1.1 \times 10^{-21}} = 3.3 \times 10^{-11}$ M

16.43 The smaller the K_{sp} value is, the lower the solubility of the substance. The three values of K_{sp} from the previous problem are 5.9×10^{-19}, 1.0×10^{-26}, and 2.0×10^{-58}. These are in decreasing order of solubility. The sulfides would be ranked FeS, SnS, and PdS from most soluble to least soluble.

16.45 $CaCO_3(s) \rightleftharpoons Ca^{2+}(aq) + CO_3^{2-}(aq)$

$K_{sp} = [Ca^{2+}][CO_3^{2-}]$

16.47 $Al^{3+}(aq) + 6H_2O(l) \rightleftharpoons Al(H_2O)_6^{3+}(aq)$

16.49 $ZnS(s) + H_2O(l) \rightleftharpoons HS^-(aq) + OH^-(aq) + Zn^{2+}(aq)$

$K = [HS^-][OH^-][Zn^{2+}]$

16.51 In an acidic solution, the value of K is much larger—10^{21} larger—than it is in water. This means it is more soluble in an acidic solution. Looking at the equation, we see that there is OH^- formed in the equilibrium. If we put this in acid, the OH^- ions will react with the H_3O^+ of the acid and cause the equilibrium of the ZnS to shift toward the products. This makes it more soluble.

16.53 For $Ca(NO_3)_2$ with a pH = 6.7.

$[H_3O^+] = 10^{-6.7} = 2.0 \times 10^{-7}$ M

For $Ca(NO_3)_2$ with a pH = 3.6, $[H_3O^+] = 10^{-3.6} = 2.5 \times 10^{-4}$ M.

A solution of $Pb(NO_3)_2$ with a pH = 3.6 would be more acidic (i.e., a lower pH) than the pH = 6.7 $Ca(NO_3)_2$ solution.

16.55 $HA(aq) + H_2O(l) \rightleftharpoons A^-(aq) + H_3O^+(aq)$

$$K_a = \frac{[A^-][H_3O^+]}{[HA]} \qquad \frac{[H_3O^+]}{[HA]} = \frac{K_a}{[A^-]}$$

16.57 Sulfuric acid, H_2SO_4, has two hydrogen ions that can come off in solution. They are independent of each other. The first one must come off before the second one can remove itself. This stepwise ionization has differing strengths.

16.59 The strong acid, let's call it HX, will ionize completely in water to produce the ions:

$HX(aq) + H_2O(l) \rightarrow H_3O^+(aq) + X^-(aq)$

There will be virtually no acid molecules in the solution, only ions; hydronium ions, and anions.

The weak acid, let's call it HA, will not completely ionize in the solution.

$HA(aq) + H_2O(l) \rightleftharpoons H_3O^+(aq) + A^-(aq)$

There will be a large amount of the original acid molecules in the solution with few ions.

CHAPTER 17

17.1 According to Table 17-1, C forms four bonds each.

a) C_2H_6 $2(e^-) + 6(1e^-) = 14e^-$

```
    H  H
    |  |
 H-C-C-H
    |  |
    H  H
```

b) C_2H_4 $2(4e^-) + 4(1e^-) = 12e^-$

The skeleton of the molecule shows us that the carbon atoms will not reach octet through single bonds. There is an extra pair of electrons that should be in the structure, according to the number of valence electrons. There will be a double bond between the carbons.

```
H       H
 \     /
  C=C
 /     \
H       H
```

c) C_2H_2 $2(e^-) + 2(1e^-) = 10e^-$

This is similar to C_2H_4, except there is a triple bond between the carbons.

H—C≡C—H

17.3 a) $N(CH_3)_3$

```
 H           H
   \   ••   /
H-C—N—C-H
   /   |    \
 H  H-C-H  H
       |
       H
```

b) $CH_3CH_2NH_2$

```
    H  H
    |  |  ••
 H-C-C-N-H
    |  |  |
    H  H  H
```

17.5 See Table 17-1.

```
    H  H
    |  |
 H-C-C-O-H
    |  |
    H  H
```

```
    H     H
    |     |
 H-C-O-C-H
    |     |
    H     H
```

```
    H  H
    |  |
 H-C-C=O
    |  |
    H  H
```

```
    H  H
    |  |
 H-C-C-N-H
    |  |  |
    H  H  H
```

17.7

```
 H H     H          H       H
 | |    /            \     /
H-C-C=C               C---C
 |      \            / \ / \
 H       H          H   C   H
                       / \
                      H   H
```

17.9 See WebLab

17.11 Saturated are compounds with the maximum number of H-atoms on each C-atom in the skeleton. This means, no multiple bonds between C-atoms. The only structure that is saturated is

```
  H H
  | |
H-C-C-H
  | |
H-C-C-H
  | |
  H H
```

17.13 Substance (a) and substance (d) are a pair of isomers of formula $C_5H_{10}O$.

```
  H                                    H
   \   O                               |  O
H-C/    \C-H                         H-C/  \H
  |      |\H          and      H       |
H-C      C-H                   |      C-H
H/  \   /  \H               H  C     /  \H
      C                     H-C=/  \C/   
     / \                    H/     | \H
    H   H                          H
```

Substance (b) and substance (f) are a pair of isomers of formula $C_6H_{12}O$.

```
                O                        H H
               //                     H   \ /    H
           H-C                        H-C--C    /
     H H       |        H      and     |     \C-O
      \|       |        |            H-C       | \H
H-C---C---C---C---C-H                H/  \   C-H
H/         / \   |  \H                    C    \H
          H   H  H                       / \
                                        H   H
```

Substance (c) and substance (e) are a pair of isomers of formula $C_4H_{10}O$.

```
                                              H
                                               \
H      H H          H                   H       O      H
 \      \|          |                    \     /       |
H-C      C          C-H     and     H     C           C-H
 /  \   /  \      /   \              \   / \   C    /   \
H    C       O         H           H-C       \/ \       H
    / \                             /         C
   H   H                           H         / \
                                            H   H
```

17.15 a)

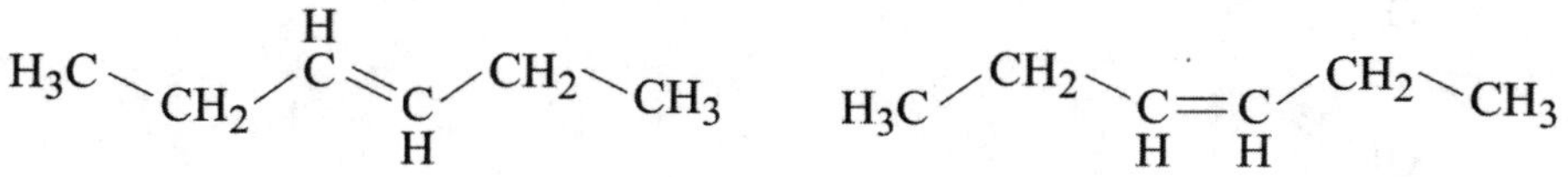

$H_3C-CH_2-CH=CH-CH_2-CH_3$ (trans) $\quad$ $H_3C-CH_2-CH=CH-CH_2-CH_3$ (cis)

b)

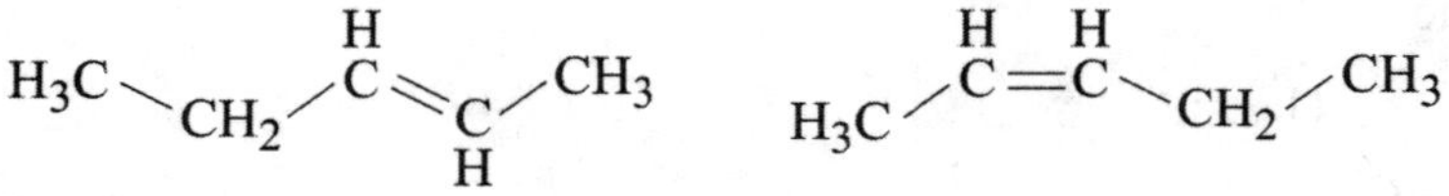

$H_3C-CH_2-CH=CH-CH_3$ (trans) $\quad$ $H_3C-CH=CH-CH_2-CH_3$ (cis)

c)

$CH_3-HC=N-CH_3$ $\quad$ $H_3C-CH=N-CH_3$

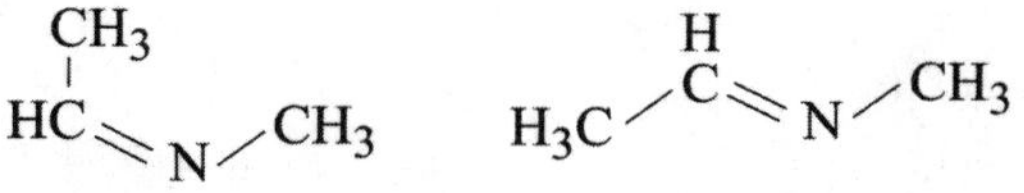

17.17 See Figure 17-13.

This might be similar to rotten fish.

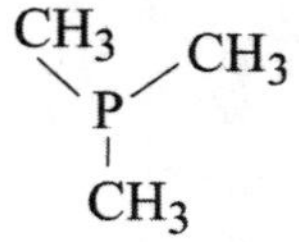

$P(CH_3)_3$

This might be of a sour or unpleasant odor.

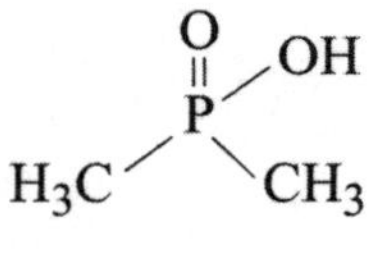

$(CH_3)_2P(=O)OH$

17.19 See Table 17-3

$HO-P(=O)(OH)-O-P(=O)(OH)-O-P(=O)(OH)-OCH_2$ (ribose, OH, OH) – N (adenine: HC, N, C, C, N, CH, N, NH_2)

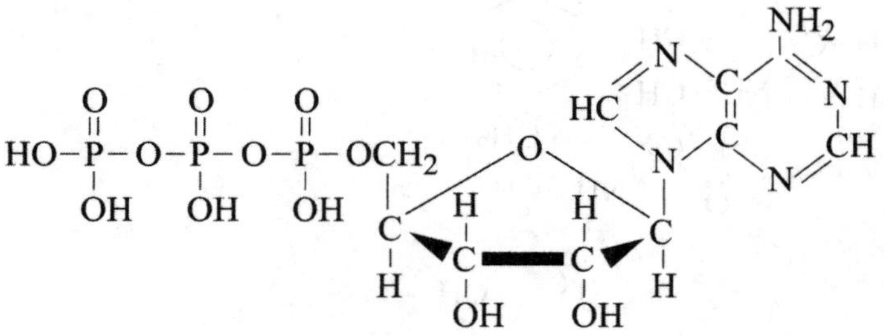

17.21 Alanine

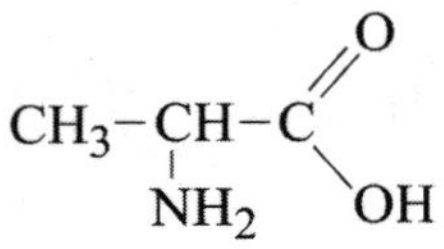

$CH_3-CH(NH_2)-C(=O)OH$

Phenylalanine

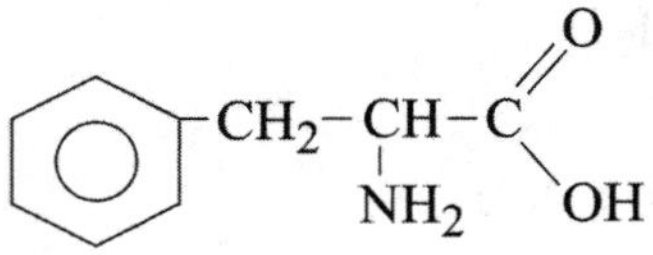

$C_6H_5-CH_2-CH(NH_2)-C(=O)OH$

Tyrosine

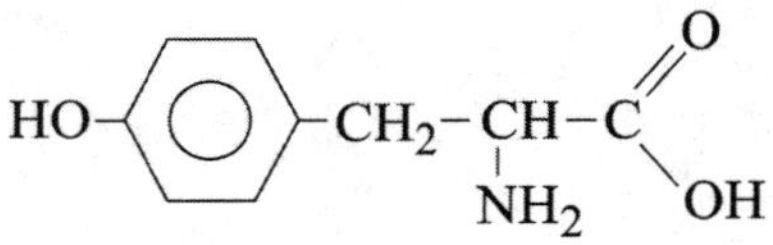

$HO-C_6H_4-CH_2-CH(NH_2)-C(=O)OH$

Peptide #1

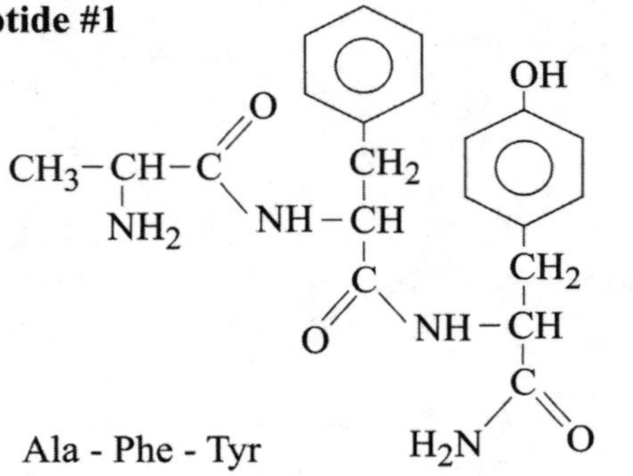

$CH_3-CH(NH_2)-C(=O)-NH-CH(CH_2C_6H_5)-C(=O)-NH-CH(CH_2C_6H_4OH)-C(=O)NH_2$

Ala - Phe - Tyr

Peptide #2

$HO-C_6H_4-CH_2-CH(NH_2)-C(=O)-NH-CH(CH_3)-C(=O)-NH-CH(CH_2C_6H_5)-C(=O)OH$

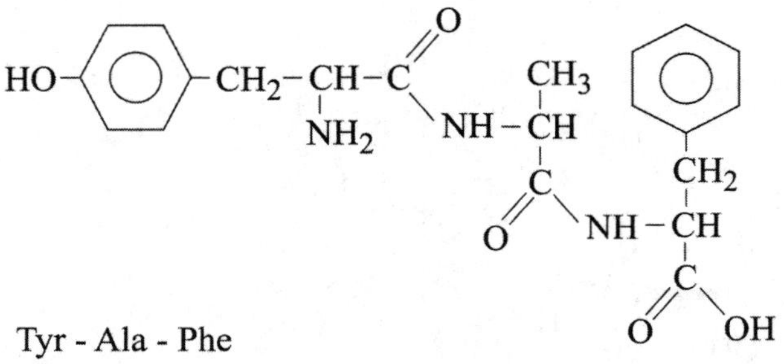

Tyr - Ala - Phe

Peptide #3

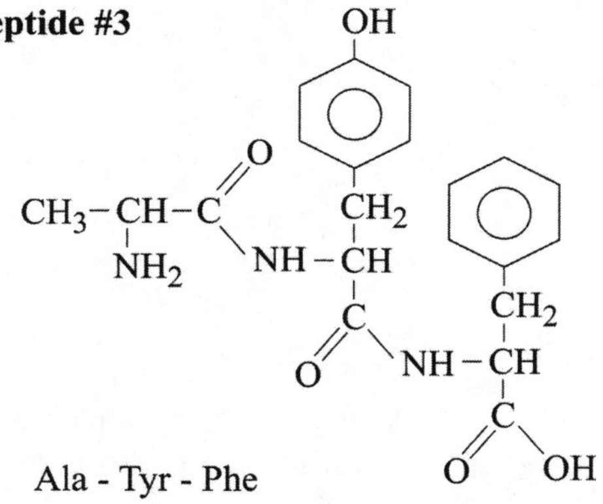

```
                        OH
                        |
                     (benzene)
              O         |
             //        CH2     (benzene)
CH3-CH-C               |         |
    |    \NH - CH              |
    NH2             |           CH2
                    C           |
                  //  \NH - CH
                  O            |
                               C
                             //  \
Ala - Tyr - Phe             O     OH
```

17.23 Find the peptide bonds $-\overset{O}{\overset{||}{C}}-\overset{H}{N}-$ to determine how many amino acids there are.

This is a pentapeptide. See Table 17-3 to identify each of the amino acids. The sequence is Asp-Phe-Gly-Ser-Cys.

17.25

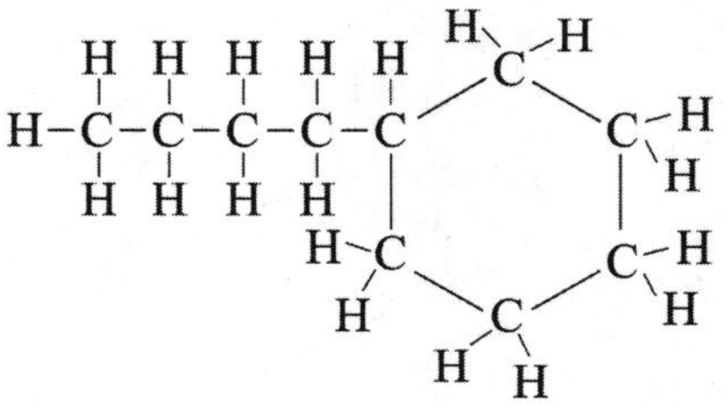

```
                      H  H
 H H H H H             \C/
 | | | | |            /   \
H-C-C-C-C-C               C-H
 | | | |   |              | \H
 H H H H   |              |
         H-C              C-H
          /  \           /  \
         H     \   C   /     H
                  / \
                 H   H
```

17.27

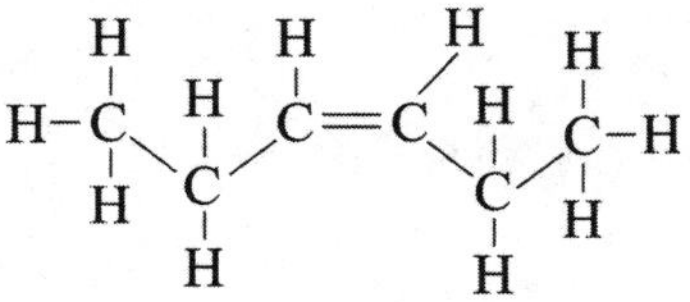

```
  H        H      H     H
  |   H    |     /  H   |
H-C   |    C=C      |   C-H
  |  \C  /    \ C  /    |
  H   |         |       H
      H         H
```

17.29 Proline

```
          O
         //
 [ring]-C
  \N/     \
   H       OH
```

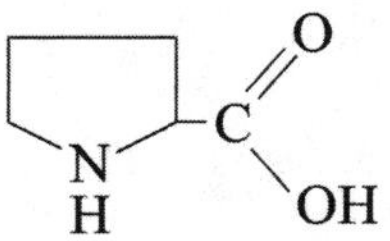

Glutamic acid

```
   O                       O
   ||                     //
   C-CH2 - CH2 - CH - C
  /                |       \
HO                 NH2      OH
```

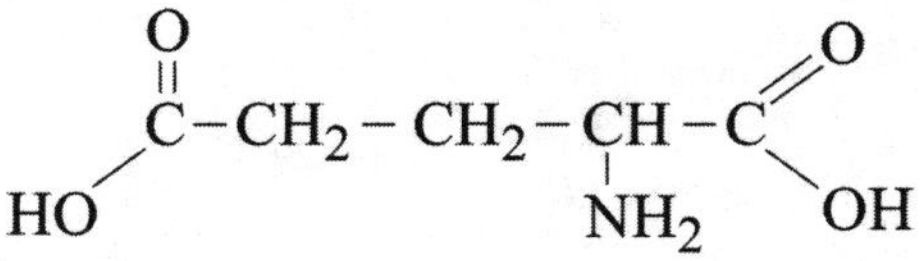

Pro - Glu

O HO O
C
C NH — CH
N
H
CH_2
CH_2
C
HO O

Glu - Pro

O O
$C-CH_2-CH_2-CH-C$
HO NH_2 N
C
HO O

GLOSSARY

absolute zero The lowest possible temperature is called absolute zero. On absolute temperature scales like the Kelvin scale, absolute zero has a value of 0.
accuracy Accuracy means that the experimental value agrees closely with some known value.
acid-base equilibria An acid-base equilibrium system contains an acid and its conjugate base.
acidic This adjective can be applied to a solution having the recognized properties of acidity, including characteristic color changes in indicators, corrosive character with metals and carbonates, and a pH less than 7. It can also be applied to a substance that can cause a solution to become acidic.
actual yield The amount of product actually isolated from a reaction.
alpha particle An alpha particle is a helium nucleus, ^4He, and is given off by an unstable nucleus in a process called alpha emission.
amino acid Biologically important molecules that form the backbone of proteins.
amphoteric A species capable of acting either as a Brønsted-Lowry acid or base, depending on the other reactants in the solution.
angular node A kind of node defined by an angle relative to the position of the atom. This gives rise to the shapes of atomic orbitals.
aqueous solution An aqueous solution is one in which the solvent is water.
arrhenius acid A substance capable of donating hydrogen ion to water to cause an acidic solution, or capable of neutralizing a basic solution.
arrhenius base A substance capable of forming hydroxide ion in water to form a basic solution, or capable of neutralizing an acidic solution.
atmosphere The atmosphere is a common unit for pressure in chemistry.
atmospheric pressure Atmospheric pressure is the pressure or force of all the gases in the atmosphere on a particular system of interest.
atom An atom is the smallest part of a substance; atoms cannot be broken down into smaller components by ordinary chemical means.
atomic mass The atomic mass is the sum of the masses of all protons, neutrons, and electrons contained in that atom.
atomic mass scale The atomic mass scale is a relative scale in which the atomic mass of each element is compared to the atomic mass of carbon, which is set to be exactly 12. Because the atomic mass scale compares two masses (the atomic mass of some element to the atomic mass of carbon-12), it has no units.
atomic mass unit or **amu** An atomic mass unit or "amu" is a defined unit in the atomic mass scale. The atomic mass unit is defined as exactly $^1/_{12}$ the mass of a carbon-12 atom.
atomic number An element's atomic number is an integer that uniquely identifies that element. It is also equal to the number of protons in the nucleus of all atoms of that element.
atomic orbital A state occupied by an electron in an atom. An atomic orbital has certain properties based on its quantum level.
atomic symbol The one-, two-, or three-letter code for an element. It always has the first letter capitalized in all uses.
Avogadro's number Avogadro's number (sometimes symbolized as N_o) is equal to 6.022×10^{23} objects. N_o is the number of objects in one mole. Thus, there are 6.022×10^{23} atoms C in one mole of C and

6.022×10^{23} molecules CO_2 in one mole of CO_2.
balanced chemical equation A balanced equation contains equal numbers of each element on the reactant and on the product side of the equation.
barometer A barometer is an instrument used to measure the height of a column of mercury that the atmosphere is able to support. The units of this pressure measurement are mmHg or cmHg, depending on the scale used to measure the mercury column
basic This adjective can be applied to a solution having the recognized properties of basicity, including characteristic color changes in indicators, caustic character with substances, and a pH greater than 7. It can also be applied to a substance that can cause a solution to become basic.
beta particle A beta particle is an electron, ${}_{-1}^{0}\beta$. These are given off by an unstable nucleus in a process called beta emission.
binary Binary refers to two of something. A binary compound contains two and only two elements.
binary compound A compound that contains two and only two elements.
biochemistry The branch of chemistry that deals with molecules and reactions important in living things.
boil When a liquid boils, it is turning into a vapor accompanied by the formation of bubbles within the liquid.
boiling point temperature The temperature at which a substance boils/condenses, i.e., converts between the liquid and gaseous phases.
Boyle's Law Boyle's Law states that at constant temperature, the volume occupied by a specified quantity of gas is inversely proportional to the applied pressure.
Brønsted-Lowry acid A species capable of donating a hydrogen ion to another species in an acid-base reaction.
Brønsted-Lowry base A species capable of accepting a hydrogen ion from another species in an acid-base reaction.
burning A chemical reaction involving the rapid combination of a substance with oxygen.
carbohydrates Biologically important molecules with the empirical formula CH_2O.
Cartesian coordinate system The Cartesian coordinate system is a rectangular coordinate system formed by the intersection of two perpendicular lines.
change Change is defined as a new arrangement of matter.
chemical bonding The way atoms are connected to one another in a chemical substance.
chemical change A chemical change is a reaction that transforms (changes) a substance into a new substance or substances with different properties.
chemical counting The application of counting rules to the counting of atoms, molecules, and formula units.
chemical element Chemical elements are the simplest kind of chemical substance, which cannot be broken down into any parts. All the atoms of an element are the same.
chemical equation A statement, in words or in symbols, that describes a chemical reaction. Names (or formulas) of all reactants and products, together with their physical states of matter, are included in the equation.
chemical formula The chemical formula indicates the elements that compose a chemical substance.
chemical property A chemical property relates a characteristic about how a chemical substance engages in chemical change, where it is transformed into a new substance.
chemical reaction A chemical reaction is a process whereby the atoms present in reactant substances are rearranged into different substances called products.
chemical stoichiometry The measurement of the components of a chemical reaction.
closed shell A situation where the structure of the electrons in an atom is full in terms of the available electron shells.
coefficients Coefficients are simple, whole numbers written in front of chemical formulas in a balanced chemical equation. Coefficients must be written in lowest terms.

combined gas law The combined gas law is $\frac{P_1V_1}{n_1T_1} = \frac{P_2V_2}{n_2T_2}$. This equation is useful whenever there is a change of one or more variables for a gas sample.
compound A compound is a chemical substance that is formed from more than one element.
concentration The concentration of a solution tells "how much solute" is dissolved in "how much solvent."
condense When a vapor turns into either a liquid or a solid, it is said to condense.
conjugate acid-base pair Two species related to one another by the presence of a hydrogen ion (in the acid part of the pair) or its absence (in the base part or the pair).
decomposition reaction When a single reaction forms two or more substances in a chemical reaction.
density Density is a physical property determined by dividing the mass of a substance by its volume. Density is an intensive property that can be used (with other criteria) to identify a substance.
dependent variable In an experiment we follow the effect of changes in an independent variable through changes in the dependent variable.
deuterium Deuterium is an isotope of hydrogen that has a mass number of 2. Its symbol is ^2H.
diatomic A diatomic group of atoms contains only two atoms; usually refers to a molecule with only two atoms.
displacement reaction A displacement reaction is a reaction in which an element is "displaced" from a compound by a second element. In these reactions, metals displace metals and nonmetals displace nonmetals.
dissociation Dissociation means "to break away from." When applied to the solution process, dissociation refers to the ionization that occurs when an ionic compound dissolves in water. Thus, NaCl dissociates into two ions, one Na^+ ion and one Cl^- ion, while $MgCl_2$ dissociates into three ions, one Mg^{2+} ion and two Cl^- ions.
electron An atom's electrons are light, negatively charged particles that surround the atom's nucleus.
electron configuration The assignments of electrons to particular principal quantum numbers and subshell designations in an atom.
electron domain The region of space around an atom occupied by a lone pair or a chemical bond to another atom (the bond may be single, double, or triple).
empirical formula The empirical formula of a substance contains the elements present in the smallest, whole number ratio to one another.
endothermic This refers to cases where a part of the universe takes heat in. In these cases q for this part of the system is greater than zero.
energy The ability to do work or produce heat. Energy comes in many different forms, and energy conversion involves changing energy from one form to another.
equilibrium A chemical equilibrium system is a dynamic system in which reactants continue to convert into products while the products also convert back into reactants. At any given time, then, all reactants and products are present in amounts determined by the rates of the forward and the reverse reactions.
equilibrium constant The equilibrium constant, K, is the numerical value of the equilibrium expression. K is unitless and valid at the particular temperature at which it is measured.
equilibrium expression An equilibrium expression is an equation in which the ratio of product concentrations to reactant concentrations is equal to a constant value at a given temperature. The ratio gives the product of the concentrations of the products divided by the product of the concentrations of the reactants, each concentration raised to a power equal to its coefficient in the balanced chemical equation.
estimated number We get an estimated number when we measure something and need to estimate part of our answer.

evaporate Either a liquid or a solid is said to evaporate when it forms a vapor.
exact number An exact number results when we count something or when a unit is defined exactly, either by complete counting or by definition.
excess reactant The reactant or reactants that are left over when all of the limiting reactant has been consumed.
exothermic This refers to cases where a part of the universe releases heat. In this case, the heat change q is less than zero.
exponent The exponent is a superscript number that indicates the number of times the base appears as a factor.
exponential functions Mathematical functions of the form $y = a^x$, where $a > 0$.
extensive property An extensive property is one that depends on the size of the sample. The measured value of an extensive property will change as the size of the sample changes.
first ionization energy The smallest energy required to completely remove an electron from a neutral atom in the gas phase.
formula mass The sum of the masses of all elements in the formula of the substance.
freeze When something freezes, a liquid becomes a solid.
functional group A special arrangement of atoms in an organic molecule, usually associated with multiple bonding or a heteroatom.
gamma ray A gamma ray is a high-energy photon given off by radioactive decay processes.
geometric isomers Isomers that differ in the arrangement of atoms around a point.
group A group of the periodic table corresponds to a single column. The groups are numbered from left to right.
heat Heat is associated with thermal energy; it is usually connected to temperature changes.
heat capacity This refers to the response of a substance to an input of heat. The heat capacity is the constant of proportionality between heat and temperature change.
heat of fusion This is the quantity of heat needed to melt a substance to a liquid. It is usually expressed in units of energy "per mole."
heat of vaporization This is the value of the heat needed to vaporize a liquid substance to a gas. It is usually expressed in units of energy "per mole."
heterogeneous Heterogeneous samples have different regions apparent to observers.
homogeneous Homogeneous samples have a uniform appearance throughout the sample.
ideal gas law The ideal gas law equation is $PV = nRT$. This equation holds strictly only for hypothetical or ideal gases but real gases can approach ideal conditions at low pressures.
independent variable In an experiment, we vary a quantity and follow the change in another variable. The variable that is varied under our control is the independent variable.
indicators Substances or materials that have characteristic color changes when the acid-base properties of a solution change. Examples include litmus paper, phenolphthalein, and bromocresol green.
inner transition elements The elements in the two sections of the periodic table usually separated out to below the table. These include fourteen elements in each case.
insoluble A substance that does not dissolve in a solvent.
intensive property An intensive property is a property that is intrinsic to a substance. The measured value of an intensive property does *not* depend on the size of the sample.
ionic bond A kind of bonding that involves the very strong forces associated with charged atoms and molecular ions. In this case, a positive charge and a negative charge are present, and the charges attract one another to form an ionic bond.
ionic compounds Compounds that contain positive and negative ions in an extended structure.
ionic solution equilibria Ionic solution equilibria are equilibrium systems containing ions in solution. Three important types

of ionic solution equilibria are: metal-ligand equilibria, acid-base equilibria, and solubility equilibria.
isomers Substances that have the same chemical formula are isomers of each other.
isotope An isotope is an atom of the same element that has a different mass number because of a difference in the numbers of neutrons in its nucleus.
isotopic mass Isotopic mass is the mass of one of the isotopes of an element.
joule This is the basic unit of energy, including thermal energy and heat.
Kelvin temperature scale The Kelvin temperature scale is a scale based on the expansion properties of gases. It is related to the Celsius temperature scale

by $T = t \times \frac{1K}{1°C} + 273.15$, where

T = temperature in Kelvin and t = temperature in degrees Celsius. A Kelvin is the same size as a degree Celsius.
leading zeroes Leading zeroes are zeroes that appear as placeholders at the beginning of a number that is between 1 and −1.
Lewis dot A representation of an electron in an atom or molecule by a dot that represents one electron.
limiting reactant The reactant that produces the smallest amount of product in a chemical reaction.
line spectrum A physical property of an element where certain wavelengths of light are emitted from an atom.
logarithmic functions Mathematical functions of the form $x = \log_a y$, where a is a positive number.
macroscopic A scale used to describe and count large samples of matter. This scale is used when we describe a chemical reaction in terms of moles of reactants and products.
magnitude The "magnitude" of a number is the size or extent of the thing measured and is correlated with the exponent of 10 when the number is written in scientific notation.
mass Mass is the measure of the amount of matter present.
mass number The mass number is the sum of the number of neutrons and protons in an atom. It can be different for different atoms of the same element, because the number of neutrons in the atom can vary.
matter Matter is anything that occupies space and has mass.
measured numbers The numbers that result when a measurement is taken.
melt When something melts we see a change in a phase when a solid becomes a liquid.
melting-point temperature The temperature at which a substance melts/freezes, i.e., converts between the solid to the liquid phase.
metal-ligand equilibrium A metal-ligand equilibrium system is composed of a metal ion, a ligand, and a complex formed of the two; a ligand is another ion such as Cl^- or CN^- or a neutral molecule such as H_2O or NH_3. The charge on the complex formed is determined by its component parts.
metric prefix A metric prefix is used with basic units such as gram, meter, and liter to indicate the magnitude of the measurement. These prefixes, which should be memorized, are listed in Table 7-5.
metric system The metric system is the numerical system of choice for scientists. It is based on powers of ten.
metric system of units The metric system of units is an international system of units used in most countries of the world and in all areas of science. Metric units are called the System Internationale d' Unites (commonly referred to as SI units).
microscopic A scale used to describe and count very small samples of matter. This scale is used when we describe a chemical reaction in terms of individual atoms, molecules, and formula units.
model A "model" applies an easy to understand verbal or visual pattern to more complicated situations.
molar mass The molar mass is the mass, in grams, of one mole of a substance. The molar mass is numerically the same as the formula mass but with units of grams mol^{-1}.

molarity Molarity is a concentration unit that tells us "how many moles of solute" are dissolved in "1 liter of solution."

$$\text{Molarity} = \frac{\text{moles of solute}}{\text{liters of solution}}.$$

mole A mole is a unit that represents a very large number, 602,213,700,000,000,000,000,000. This number is often called Avogadro's number. Mole is abbreviated mol.

mole ratio A fractional comparison of mole amounts of atoms, molecules, and formula units. Just as the number of atoms in a molecule or formula unit can be counted, so can the number of moles of atoms in a mole of molecules be counted. Thus, the formula $C_3H_6O_3$ can have the atom/molecule and mole/mole ratios:

$$\frac{\text{3 O atoms}}{\text{1 molecule}} \Rightarrow \frac{\text{3 moles O}}{\text{1 mole compound}}.$$

mole ratio for a chemical reaction The ratio of the number of moles of substances in a chemical reaction. The values of this ratio are taken from the stoichiometric coefficients of the balanced chemical equation.

mole ratio for a chemical substance The ratio of the number of moles of an element to the moles of a chemical substance, or of different elements in the substance.

molecular formula The molecular formula of a substance has a mass equal to the molar mass of that substance. It is either the empirical formula or some whole number multiple of the empirical formula.

molecular substance A substance made up of molecules as its basic building blocks.

molecule An arrangement of atoms with a fixed formula and set of connections.

natural abundance Isotopes of an element are present in a random sample of that element in particular amounts that add to 1 and are relative to one another. This amount is called the natural abundance of that isotope.

negative exponent rule Raising an expression to a negative power is the same as raising its reciprocal to the positive power. We will call this the negative exponent rule.

neutralization The reaction that occurs when acidic and basic solutions are mixed to produce a solution that has neither basic nor acidic properties.

neutron The neutron is a neutral particle in the nucleus of the atom that has a mass very close to that of the proton.

nomenclature A system of naming things by definite rules; in chemistry this includes recognizing if a substance is likely to be molecular.

nonmetal The nonmetals are hydrogen and the elements in the top right area of the periodic table. With rare exceptions, the nonmetals never exhibit metallic properties.

normal boiling point The "normal" boiling point is the boiling point temperature measured at sea level.

nucleoside One of a small number of molecules made of a nitrogen-containing base attached to a carbohydrate ring. These assemble into the DNA and RNA chains of the genetic code.

nucleus The innermost part of the atom that contains most of the mass of the atom.

organic chemistry The branch of chemistry devoted to the study of carbon-containing molecules, including their structure and reactions.

oxidation Oxidation is a process by which an atom loses electrons. Oxidation is recognized by an increase in the positive value of the atom's oxidation number.

oxidation number The oxidation number of an atom in a compound is the same as its charge if it exists as a monatomic ion. The oxidation number of an atom in a polyatomic ion is the hypothetical charge found by applying the principle of charge balance.

oxidation-reduction reaction An oxidation-reduction reaction is a reaction in which one substance is oxidized and a second substance is reduced. These reactions are recognized by changes in oxidation numbers from the reactant to the product side of an equation.

percentage yield The ratio, in percent terms, of the actual yield to the theoretical yield.

period A period of the periodic table corresponds to a single row. The periods are referred to as 1st, 2nd, 3rd, etc.

periodic law The periodic law describes how chemical elements have properties that are similar in a repeating, or periodic, fashion.

pH A logarithmic function used to describe the concentration of hydronium ion in a solution. $pH = -\log_{10}[H_3O^+]$. The inverse function relates hydronium ion to pH, $[H_3O^+] = 10^{-pH}$.

physical change Physical change occurs without changing what chemical substances are present.

physical constant A physical constant is a measurement of a physical property of a substance that does not change when measured at the same conditions of temperature and pressure.

physical property Physical properties tell us about the appearance of a substance or mixture. Each chemical substance has a unique set of physical properties.

physical state The physical state of a sample generally refers to whether it is a liquid, solid, or a gas.

pOH A logarithmic function used to describe the concentration of hydroxide ion in a solution. $pOH = -\log_{10}[OH^-]$. The inverse function relates hydronium ion to pH, $[OH^-] = 10^{-pOH.}$

polyatomic ions "Polyatomic" literally means "many atoms." A polyatomic ion is an ion that contains more than a single element.

polypeptides Molecules built from some small starting unit and assembled into very long chains, known as polymers.

polyprotic acids Species capable of losing more than one hydrogen ion in acid-base reactions. The hydrogen ions are lost in sequence, not at the same time.

polyprotic bases Species capable of accepting more than one hydrogen ion in acid-base reactions. The hydrogen ions are accepted in sequence, not at the same time.

precision Precision means that the experimental value is roughly the same every time that measurement is done. We report numbers to the number of digits consistent with the precision of the instrument.

pressure The pressure of a gas sample is a measurement of the forces exerted on or by the gas molecules in the sample.

principal quantum number The positive integer ($n = 1, 2, 3,...$) that describes a quantum level for an electron in an atom.

products Products are the elements and compounds formed in a chemical reaction.

property A property of something is a fixed characteristic of the thing, found for all samples of something.

proportion A proportion is an equation in which two ratios are set equal to each other.

proportional reasoning Proportional reasoning is a problem-solving technique based on the use of proportions. One ratio in the proportion gives some known relationship. The second ratio contains the same quantities in the numerator and denominator as the known ratio does but, in this case, one of the values is not known.

proton The proton is the positively charged particle in the nucleus of the atom.

pure chemical substance Pure chemical substances are the simplest components of mixtures.

quadratic formula The quadratic formula is $x = \dfrac{-b \pm \sqrt{b^2 - 4ac}}{2a}$, where a, b, and c represent the coefficients in the quadratic equation. It is used to solve for the two roots (x) of a quadratic equation.

quantum theory A physical theory that explains the behavior of matter in terms of particular energy levels—quantum levels—within atoms and other species.

radial node A place where an atom has no amplitude. The "radial" indicates that this node is a fixed distance from the nucleus, defining a sphere.

radiation Radiation due to radioactivity refers to the high-energy particles and energy given off by a nucleus as it seeks to achieve a more stable number of protons and neutrons.

radioactive dating A process that correlates the amount of ^{14}C remaining in a material with its age.

radioactivity Radioactivity is the spontaneous emission of particles and energy from a nucleus.
radioisotope A radioisotope is a radioactive isotope. Radioisotopes have many uses in medicine.
random error Random error is due to normal fluctuations that occur when repeating the same task a number of times.
reactants Reactants are elements and compounds that appear on the left side of a chemical equation. They are the species that react or undergo a chemical change into other elements and compounds called products.
reaction yield The amount of product in a reaction.
reduction Reduction is a process by which an atom gains electrons. Reduction is recognized by a decrease in the positive value of the atom's oxidation number.
representative elements The elements in the two "tall" sections of the periodic table are the representative elements; these comprise groups 1, 2, and 13–18 in the Arabic numbering of the groups, and I to VIII in the Roman numbering.
saturated solution A solution that cannot dissolve any more solute at that temperature and pressure.
scientific notation A number written in scientific notation is written as the product of a number, n, between 1 and 10, i.e., $1 \leq n < 10$ and a power of 10.
semimetal The semimetals are found at the border between the metals and nonmetals. They often exhibit properties, especially when pure, that are like the metals.
significant figures The significant figures in a measured number include all digits you can read with certainty plus one uncertain digit that is estimated.
slope of a line The slant of a line is called the slope of the line. We use m to indicate slope.

$$m = \frac{\text{rise}}{\text{run}} = \frac{\text{vertical change}}{\text{horizontal change}} = \frac{\text{difference in } y \text{ values}}{\text{difference in } x \text{ values}} = \frac{\Delta y}{\Delta x}$$

slope-intercept equation The slope-intercept form of a general equation is $y = mx + b$, where the slope of the line is given by m and the y-intercept is given by b.
solubility The maximum amount of solute that dissolves in a solvent at a specified temperature and pressure.
solubility equilibria A solubility equilibrium contains a compound that is present as a solid and in solution.
soluble A substance that dissolves in a solvent.
solute The solute is the substance that is dissolved to make a solution.
solution A solution is a homogeneous mixture of two or more substances.
solvent The solvent is the substance in which the solute is dissolved. Together, solute and solvent make up a solution.
specific heat capacity This is a heat capacity for a substance per one gram of the substance. It is an intensive property of that substance.
standard pressure Standard pressure is defined as the pressure of the atmosphere exerted on a column of mercury when measured at sea level. 1 atm = 760 mmHg = 760 torr = standard pressure
stereoisomers Isomers that differ based on the arrangements of atoms around a point.
stoichiometric amount An amount of a substance that is exactly equal to the amount predicted by the stoichiometry of the reaction.
STP STP is the abbreviation for standard temperature and pressure. STP for gases is 273.15 K and 1 atmosphere pressure.
strong acid A strong acid ionizes essentially all of its hydrogen ions in water solution.
strong base An Arrhenius base that completely reacts in water, usually involving the complete dissolution of a hydroxide salt.
structural isomers Isomers that differ in the order of atoms.
sublimation Sublimation is the process when a solid turns into a vapor.
subshell The s, p, d, subdivisions of the principal quantum levels. In electrons with

two or more electrons these subshells are different for the same principal quantum number.
surroundings The part of the universe that is not under direct study.
synthesis reaction A synthesis (putting together) reaction is a reaction in which two or more substances combine to give a new compound.
system A system is the part of the universe that is under study. It is often isolated from the surroundings.
systematic error Systematic error may be due to some fault in the lab equipment such as a poorly calibrated thermometer, a malfunctioning spectrophotometer, or contamination in a standard solution.
systematic name A name given to a substance according to a general naming system.
the principle of charge balance Charge balance means that the sum of the charges on the anion must be equal and opposite to the charges on the anions: charges on cations + charges on anions = 0 or charges on anions = –charges on cations.
theoretical yield The amount of product predicted by stoichiometry.
trailing zeroes The zero at the end of this number is called a trailing zero because of its position at the end of the number.
transition element The elements in the central portion of the periodic table, comprising groups 3–12 in the Arabic numbering of the groups.
tritium Tritium is an isotope of hydrogen that has a mass number of 3. Its symbol is ^3H.
trivial name A nonsystematic name given to a substance; a trivial name is often the most common way to refer to certain substances, like water.
unsaturated solution A solution that is able to dissolve more solute at that temperature and pressure.
valence electron A valence electron in an atom is among the outermost and most accessible electrons in the atom. Valence electrons are involved in the atom-atom interactions most important in chemistry.
vapor Vapor is a name given to the gas formed when a liquid evaporates.
weak acid A weak acid only partially ionizes its hydrogen ions in water solution. This results in an equilibrium state.
weight Weight is a measurement of the amount of matter present in a sample as it is acted upon by gravity.
x-intercept The x-intercept of the line is the point at which the line crosses the x-axis. To find the x-intercept, let the value of y equal zero and solve for x.
y-intercept The y-intercept of the line is the point at which the line crosses the y-axis. To find the y-intercept, let the value of $x = 0$ and solve for y.
zero power rule Any non-zero expression raised to the zero power is equal to 1.